Wind Power

The wind power business has grown from a niche sector within the energy industry to a global industry that attracts substantial investment. In Europe wind has become the biggest source of new power generation capacity, while also successfully competing with the gas, coal and nuclear sectors in China and the US.

Wind Power looks at the nations, companies and people fighting for control of one of the world's fastest growing new industries and how we can harness one of the planet's most powerful energy resources. The book examines the challenges the sector faces as it competes for influence and investment with the fossil fuel industry across the globe. Over the course of this volume, Backwell analyses the industry players, the investment trends and the technological advancements that will define the future of wind energy. This second edition is revised throughout and contains new material on frontier wind markets and industry consolidation, as well as the cost reductions and market gains that led to 2015 being a landmark year for the big wind turbine companies.

This is an important resource for professionals working in wind and wider renewable industries, energy finance, conventional energy companies and government as well as researchers, students, journalists and the general public.

Ben Backwell has spent most of his career covering international energy markets and finance. He worked for international news agencies and as an analyst in Houston, Caracas, New York, Rio de Janeiro and Buenos Aires before moving back to his native UK in 2006. He then worked as an analyst covering major oil companies before joining renewable energy news service *Recharge*, becoming Editor-in-Chief in 2012. He joined FTI Consulting as a Managing Director in 2015 after taking part in the creation of the Solution Wind advocacy campaign ahead of the COP21 climate negotiations. He holds a master's degree in Politics from the University of London.

Wind Power

The Struggle for Control of a
New Global Industry

2nd edition

Ben Backwell

Second edition published 2018
by Routledge
2 Park Square, Milton Park, Abingdon, Oxon, OX14 4RN

and by Routledge
711 Third Avenue, New York, NY 10017

Routledge is an imprint of the Taylor & Francis Group, an informa business

© 2018 Ben Backwell

[First edition published by Routledge 2014]

British Library Cataloguing-in-Publication Data
A catalogue record for this book is available from the British Library

Library of Congress Cataloging-in-Publication Data
Names: Backwell, Ben, author.Title: Wind power : the struggle for
control of a new global industry / Ben Backwell.
Description: 2nd edition. | Abingdon, Oxon ; New York, NY :
Routledge, 2018.
Identifiers: LCCN 2017027138| ISBN 9781138082410 (hb) |
ISBN 9781138082427 (pb) | ISBN 9781315112534 (ebk)
Subjects: LCSH: Wind power industry. | Renewable energy sources. |
Energy development.Classification: LCC HD9502.5.W552 B33 2018 |
DDC 333.9/2--dc23LC record available at https://lccn.loc.gov/2017027138

ISBN: 978-1-138-08241-0 (hbk)
ISBN: 978-1-138-08242-7 (pbk)
ISBN: 978-1-315-11253-4 (ebk)

Typeset in Goudy
by Fish Books Ltd.

Contents

Illustrations

Figures

Tables

Foreword

Wind power is the first of the so-called 'new renewables' to establish itself as a mainstream power source. Utility-scale wind projects operate in more than 90 countries, with 29 countries having more than 1,000 MW installed, supplying more than 4 per cent of the global electricity supply. Total installations at the end of this year will be well over 500 GW, and recent projections show that wind could be supplying 14–20 per cent of global electricity by 2030.

Since a small company built the world's first commercial wind farm at Crotched Mountain near my childhood home in southern New Hampshire in late 1980, the industry has transformed itself beyond all recognition. It took 18 years to get to the level of 10,000 MW in 1998, and another 10 years to reach 100 GW in 2008. Passing 200 GW in early 2011, and 300 GW by the end of 2013, installed capacity has now passed 500 GW, reaching double-digit penetration of electricity supply in Denmark, Uruguay, Ireland, Portugal, Spain, Germany, and 14 US states. From a few small entrepreneurs in the US and Denmark, wind energy has emerged as a technology embraced by most of the world's major energy companies and utilities. Wind power supplies increasingly cost-competitive, carbon-free electricity in an ever-expanding number of markets around the world, enhancing energy security, stabilising electricity prices and creating good-quality jobs.

Ben's book is a valiant attempt to nail down a snapshot of a dynamic industry reaching maturity and going global while at the same time being buffeted by myriad forces beyond its control. While no 'insider' will agree 100 per cent with every bit of his analysis, most of it rings pretty true to me. He explores all the main themes while pointing out that what the future will bring is still very much in play, and importantly, draws the fundamental link between climate policy and the development of the industry, which is not often apparent in the daily to and fro of the energy policy debate. In fact, it is this disconnect between the climate and energy policy debate both nationally and internationally that is responsible for the start–stop nature of support for wind in many key markets.

At the same time, of course, wind has established itself as a mainstream power source on its own merits as a supplier of affordable, clean and indigenous energy; even in the face of massive fossil and nuclear subsidies and the lack of an effective carbon price. He makes the case very clearly that wind's greatest enemy is not its variable production, but the volatility of policy swings and roundabouts that stand in the way of delivering the maximum quantity of carbon-free electricity at the lowest possible price.

I strongly recommend Ben's book to those who want to get a broad picture of just how far wind energy has come in the past three decades, as well as the promise and challenges of the road that lies ahead.

Steve Sawyer, General Secretary, Global Wind Energy Council (GWEC)

Acknowledgements

Taking part in the growth of the wind industry, after years spent primarily covering fossil-fuel energy and finance, has been an amazing experience. As befits an industry that is still relatively young and – more importantly – one whose participants actually feel they are doing something worthwhile, I have been impressed since day one by the amount of openness in the wind industry. The idealism and pioneering spirit that is in the DNA of wind can still be felt even in the biggest corporate players in the sector.

Over the last eight years, I have interviewed and worked with some remarkable people, from some of the founders and early pioneers to long-term advocates, entrepreneurs and the tenacious engineers who, day by day, are out there finding new solutions to make the technology bigger, smarter and more efficient.

I have been overwhelmed by the positive response I have had when I told people I was writing this book, with a whole number of senior industry figures agreeing to be interviewed, suggesting themes, and reading and checking sections and chapters. Any mistakes of course, are entirely my own.

I would like to thank in particular Henrik Stiesdal, the former CTO of Siemens Wind, one of the inventors of the modern wind-turbine industry and one of its most critical thinkers, whose feedback gave me the confidence that I was on the right track and who has supported me as a friend and mentor through the last decade.

GWEC General Secretary, Steve Sawyer, the industry's most tireless advocate, has provided constant encouragement and valuable perspective since I started working in the sector, constantly reminding the wind industry of the wider objective of fighting climate change.

Mainstream Renewable Power's founder Eddie O'Connor has been a constant source of inspiration as someone willing to take on big challenges time and time again and come out with a result. Adam Bruce, Mainstream's Head of Global Corporate Affairs has provided constant insight and new thinking on policy and regulatory frameworks, as has Vestas Vice President and GWEC Chairman Morten Dyrholm.

Christian Kjaer – who was the CEO of the European Wind Energy Association (EWEA) at a crucial period in the development of the wind industry, also contributed valuable insight and experience for the first edition of this book.

And I have been greatly helped by having access to insight from some of the most knowledgeable analysts in the industry; Aris Karcanias, co-lead of FTI Consulting's Clean Energy Practice, Feng Zhao, Robert Clover and Eddie Rae.

Alfonso Faubel – formerly of Alstom Wind and The Switch's Jukka Pukka Makinen gave invaluable insight into manufacturing systems and supply-chain management; Blade Dynamics' Pepe Carnevale and Theo Botha gave me a new perspective on composites and rotor blades; while much of what I have learnt about offshore wind is thanks to Chairman of 8.2 Aarufield and industry association Renewable UK, Julian Brown.

I have also benefited from discussions over a period of time with Andrew Garrad – another of the industry's pioneers and one of its most compelling champions; Goldwind's Wu Gang; E.ON's Michael Lewis; Enel's Francesco Starace; wind industry advocate turned Global Solar Council Chairman, Bruce Douglas; IEA Renewable Energy Head, Paolo Frankl, GE's Rob Sauven and Anne McEntee; the Crown Estate's Huub den Roodjen; Nordex-Acciona CEO José Luis Blanco; Mainstream Renewable Power's Andy Kinsella, Argentinian congressman and green energy advocate Juan Carlos Villalonga; Acciona Energía CEO Rafael Mateo, Gamesa's David Mesonero; RenewableUK CEO Hugh McNeal; Offshore Renewable Energy Catapult CEO Andrew Jamieson, and ABEEólica's Elbia Melo.

I would like to commend the critical but supportive coverage of the industry by the *Financial Times'* Pilita Clark and Andrew Ward, Bloomberg's Jessica Shankleman and Business Green's James Murray. Bloomberg New Energy Finance's Michael Leibreich and Angus McCrone have played a big role in getting influencers and officials around the world to wake up and smell the coffee on renewables through their hard-hitting research and data.

Special thanks go also to, Malgosia Bartosik, Stewart Mullin, Oliver Loenker, Michael Zarin, Bart Doyle, Juan Guillermo Walker, Pete Clusky, Sarah Merrick, Ramon Fiestas, Emmet Curley, Juan Diego Diaz, Klaus Rave, Rob Hastings, Ursula Guerra, Sonia Franco, Nadia Weekes, Shaun Campbell, David Weston, Sebastían Kind, Michael Holm, Morten Albaek, Stephen Thomas, Eduard Sala de Vedruna, Frederick Hendrik, Enrique de las Morenas, Anders Søe-Jensen, Ronnie Bonnar, Per Krogsgaard, Birger Madsen, Vicente Trullench, Jean Felber, Jennifer Webber, Gordon Edge, Ramesh Kymal, Michael Holm, Garth Halliday, José Antonio Miranda, Robin Palao Bastardés, Heikki Willstedt, Jayasura Francis, Isabelle Prosser, Rodrigo Ferreira, Luis Adao da Fonseca, Vineeth Vijayaraghavan, and Jonathan Collings.

I would also like to thank FTI Consulting for supporting my efforts in writing a revised second edition of this book, and in particular, John Waples, Deborah Scott, Alaric Marsden, Ed Westropp, Ben Brewerton, Francesca Boothby, Emerson Clarke, Caroline Cutler, Daniel Hamilton, Amy Yiannitsarou, and Miranda Bray.

Finally, thanks to my wife Melissa, and my two sons Dimitri and Thierry.

Introduction

The struggle for the global wind-power market

From the arid plains of Gansu province in China to Oklahoma, to the rough waters of the North Sea, wind power is on the march.

Once a small industry selling mainly to farmers and cooperatives, and dominated by a couple of small, mainly Danish firms, the sector has become a key investment area for industrial heavyweights such as Siemens, General Electric, and Mitsubishi, in the search for zero-emission power generation.

In key markets in Europe, the United States and China, wind has gone from being a niche power source, to one that is increasingly challenging 'conventional' energy (coal- and gas-fired thermal generation and nuclear), and along the way disrupting the traditional business model of the world's big power utilities. In 2012, wind became the biggest source of new power generation in the US for the first time ever – with a record 13.2 GW of new installations – and it had already won this place within the EU back in 2008.

Key to the industry's rise have been government incentives and targets around the global attempt to limit carbon emissions, including landmark renewable-energy targets in the EU, national feed-in tariff and green certificate schemes, state portfolio standards and the production tax credit (PTC) in the US. In some major European countries, large-scale wind power – and in particular the growing offshore sector – is practically the only way to reach their 2020 targets and close the gap in capacity caused by the decommissioning of older coal, gas and nuclear plants; while for the US and China, wind is essential if the two countries want to make any progress on stemming their rise in carbon emissions.

However, government support has not been the whole story and is set to become a far less important part of the picture going forward.

The impact of the March 2011 accident at the Fukushima Daiichi nuclear plant in Japan and the subsequent decision by Germany and other nations to shut down or put on hold their nuclear industries, strongly added to wind's momentum. In many countries, a lack of available fossil fuels – and the high cost of importing them – makes wind a logical way to add domestically produced power quickly, cheaply and cleanly, while creating local jobs. Politicians in India are fast waking up to the fact that cost and infrastructure bottlenecks make building new coal plants a pointless exercise; while in China, growing public pressure over air pollution means that policymakers have steadily increased their ambitions for wind as they look for alternatives to coal. Even though it was short on legally binding targets, the Paris Agreement of December 2015 seemed to enshrine a new level of international consensus over climate change and a determination to make faster progress in combatting CO_2 emissions.

But perhaps the most significant development since the first edition of this book was published has been dramatic falls in costs for wind power, driven by scale, supply chain efficiencies and technology innovation. Once seen by traditional voices within government and the power industry as too costly, wind has shown in a growing number of places, from Texas to the Brazilian North East to South Africa and Chile that it can compete on price with conventional power and displace incumbent fossil power sources through open competition.

In Brazil, record capacity factors have allowed wind-power projects to win power contracts at prices as low as around US$42/MWh, while gas producers were unwilling to take part in a government power tender with a floor price of R$140/MWh (US$67/MWh). [All conversions are approximate and were calculated using the conversion rate at the time of writing].

In the US, the wind industry has shown itself remarkably resistant to lower gas prices, despite all the fanfare over the 'shale gas revolution'. The sector has seen record growth in the face of Henry Hub natural gas prices at lows of US$2–3/mn BTU.

In South Africa, wind-power companies have bid at prices that are the equivalent of one half of the cost for new-build coal. And in Chile's landmark power tender in 2016, companies like Mainstream Renewable Power and Enel won the lion's share of capacity available with prices for wind as low as US$38.8/MWh, beating incumbents' already built fossil plants on price.

As we shall see, the move to competitive auctions is producing dramatic falls in wind-power costs in Europe too, and this is set to create a new inflection point for policy makers, as the case for investment and new-build nuclear and fossil becomes more tenuous.

Moving into the mainstream brings new challenges, however. As Angela Merkel told Germany's Bundestag just before her re-election as Chancellor in September 2013, renewables no longer occupy a niche, 'but are part of the overall generation mix', adding 'that leads to entirely new problems'.

As wind's weight in the generation mix has grown, so have the challenges grown in integrating their output in the grid, and managing their effect on the traditional utilities that dominate in both power generation and distribution in most geographies.

Power-system operators and government officials have raised opposition to wind – and solar – due to the effects of large-scale amounts of renewable power on system balancing.

Officials use the term 'intermittency' to describe wind power; they argue that traditional 'base-load' generation is needed, and that renewables create the need for parallel investments in back-up power, which makes them prohibitively expensive (see the front cover story in the *Economist*, 'Clean Energy's Dirty Secret', 25 February 2017).

It is true that in some places muddled regulation and market design have meant that large-scale deployment of renewables has had perverse effects in the short term. Germany's *Energiewende*, for example, has seen an impressive amount of renewables deployed, as well as a comeback for the dirtiest forms of brown coal generation, leading to a net rise in emissions for several years.

Utilities like Iberdrola, ENEL, RWE and E.ON have been key investors in wind energy, but they have found that the growth of the business has increasingly affected their traditional businesses, pushing mostly gas-fired power generation off the grid when the wind is blowing, for instance.

However, the consensus among both grid operators and utilities is that change is unstoppable and needn't cause an increase in costs. While utilities have often continued to lobby for their fossil-fuel generation business – the call for capacity markets in some European countries, for example – they are taking a series of steps to manage the transition, including continued investment in wind and key enabling technologies such as intelligent grids and storage, and carrying out 'corporate splits' between their clean power and legacy fossil businesses to create more visibility for their shareholders on their future business models.

Policymakers and regulators meanwhile, generally recognise that instead of slowing down renewables, investments are needed in the long-term solutions that are necessary for the full modernisation of power systems – cross-border power markets, a fully interconnected international grid (the 'supergrid'), storage technologies, commercial models for demand response, and smart metering.

Meanwhile, public opinion has remained resolutely positive about wind, despite politicians sensing that renewables can be an easy scapegoat to blame for unsustainably expensive power systems, and the activities of a section of the conservative press that border on the pathological. A series of hysterical campaigns, which make all sorts of claims (for example, that turbines are inefficient, that they blight the landscape, or even that they make you ill), have spectacularly failed to change the public's gut feeling that power generated from an entirely natural and free source is a good idea.

We have come a long way since the first groups of inventors and hobbyists who started the modern wind industry. The first chapter in this book looks at the creation of the world's wind market and the rise of the mainly European wind companies – led by Denmark's Vestas – to global status. Chapter 2 looks at the entry of big global engineering giants – like Germany's Siemens and the US's GE – into the business as it gained momentum, and the entry into the business of the big power utilities, which depending on who you speak to, have been key promoters of and key barriers to the growth of the industry. It also looks at the growth in the sector in the US, which, although hardly a hawk in terms of climate negotiations, has consistently been the world's single largest market for renewable energy – until the rise of China, which we tackle in Chapter 3. In industrial terms, the battle between European, US and Chinese players is one of the key themes of this book.

Following China, in Chapter 4 we look at some of the big new players, namely Brazil, where the industry is seeing record capacity factors, rock-bottom prices and breakneck growth; and India, whose current power mix and growing economy gives it enormous potential. Arguably, it will be the success or failure of wind in defining the energy economies of these countries that will decide whether the struggle to limit carbon emissions can be successful.

We then look at the fast-growing offshore sector in Chapter 5. Over the last decade, offshore wind has become one of the fastest growing areas within the wind industry. From small beginnings in a couple of pilot programmes in Denmark and the UK, offshore wind now constitutes one of the greatest engineering challenges of our time, as projects grow to a massive scale. The largest of the UK's already constructed offshore wind farms has a capacity of 630MW, while another with 1.2GW capacity has reached financial close and is due to be commissioned in 2020. Turbines now grow to skyscraper size and wind farms move out into water depths of up to 70m. Offshore wind costs are still significantly higher than for onshore wind; and yet there are strategic factors pushing forward the growth of the sector in places like the UK, where planning and space constraints on land combined with legally binding emission reduction targets and the exorbitant costs of building nuclear power create what is virtually an imperative to build offshore. The industry has long predicted that reliability improvements, scale and growing expertise in the installation process would push offshore's costs steadily downwards. Now, these predictions are becoming vindicated dramatically in a series of long-term power tenders in the Netherlands, Denmark, Germany and the UK.

And while the vast bulk of offshore wind is currently in the waters of the UK, Germany and Denmark, other countries, notably China and Japan, have the capacity and the incentive to grow their offshore sectors aggressively, something which can already be attested by the appearance of a series of new Chinese mega turbines, and Japan's Mitsubishi Heavy Industry's formation of its offshore joint venture with Vestas.

In Chapter 6, we see how the precipitous nature of wind's growth into a major global manufacturing left it ill prepared for the aftermath of the disastrous Copenhagen climate summit in December 2009. The failure of the talks on the home ground of the world's most hawkish countries in terms of taking action against emissions was a profound shock to many who took part, including the writer of this book.

Taking their signal from the top, sceptical politicians and incumbent energy lobbies stepped up criticism of the climate agenda and renewables. Wind was no longer the sexy industry on the front cover of *Forbes* magazine. The failure of talks on a binding climate agreement was not the only problem, of course. The Copenhagen fiasco also coincided with the effects of the subprime crisis being felt in earnest. As a capital-intensive sector, the liquidity squeeze hit wind energy hard compared to other sectors. Commercial financing for wind farms in Western markets largely dried up, while the utilities that had been largely driving expansion began to feel their balance sheets tighten. And, as the subprime crisis began to morph into the Eurozone crisis, key areas for wind expansion such as Spain and Portugal slammed on the brakes, while others began to slow down. Coupled with uncertainty over government support, the arrival of cheap shale gas in the US, and the onset of massive grid curtailment issues in China, the wind industry was facing a perfect storm.

Not that the industry stopped growing of course. But the slowdown in expansion and the continued entry of new players and manufacturing capacity into the industry meant that by 2011, some of the sector's most important players were facing huge overcapacity, bloated workforces, big debts, falling turbine prices and rock-bottom equity values. How some of the key players in the industry have coped with the post-Copenhagen hangover is the subject of Chapter 7.

Slump and near bankruptcy was replaced once again by boom in the 2014–2016 period. As I argue in Chapter 8, however, the basis for this new expansion is very different from in previous growth periods. Wind power, along with solar has reached a decisive 'tipping point' as it reaches the long sought after 'grid parity' and can be deployed as cheaply or more cheaply than fossil-fuel generation in a growing number of markets around the world.

In Chapter 9, I look at the impact of both a bigger market and more competitive prices on the wind-turbine industry. The most dramatic effect of the new conditions has been to trigger a long predicted wave of consolidation among wind-turbine and component suppliers, which has created a small number of globalised, well-capitalised giants that are likely to dominate the market in the next period.

In the final chapter of this book, I look at some of the key challenges facing the industry. These include how to deal with the volatility – much of it politically induced – that has been a historic feature of the wind market, and the challenges of running global manufacturing operations in a situation where growth often follows a pattern of stop–go cycles across different countries and geographies. Related to this is the challenge of growing the wind industry and the companies in it to a scale where a radically new cost structure would emerge, and the role that progress towards a 'mature' industry and consolidation plays within this.

Looking to the future, the wind industry faces the challenge of how to deal with its own success as a significant source of power within national and international power systems and the question of how it relates to disruption caused by advances in the technology sector. We shall argue that to be successful in tomorrow's energy market, companies will no longer be able to see themselves as simply providing a piece of equipment, but will need to take a 'whole system' approach, looking at how wind technology can be integrated into systems and deployed by customers in the most efficient way possible in conjunction with complementary technologies such as energy storage, demand response, and electric vehicles.

We also look at the impact of solar PV, which has gone from being seen as a smaller and more expensive 'sister' renewables technology to a serious commercial rival.

1   From Maoism to Learjets

Turbine makers go global

On September 27, 2013, at simultaneous early-morning press conferences in Copenhagen and Tokyo, officials from Vestas, the world's largest wind-turbine manufacturer, and Mitsubishi Heavy Industries (MHI), one of the world's biggest industrial conglomerates, announced a ground-breaking joint venture.

The joint venture, officials announced, was to be based around Vestas' planned V164 offshore wind turbine. With an 8MW capacity, the wind-turbine's rotor diameter is bigger than the London Eye at 164m, while its height at 187m is taller than the Gherkin building in the same city. The turbine is designed to be installed in the harsh conditions of the North Sea in water depths of up to 70–80 metres and to operate for 25 years continuously.

The offshore wind business is still relatively young and there are many uncertainties. Project developers and investors prefer to deal with companies that have the financial strength to see through their warranties in case of any large-scale equipment failure. By teaming up with MHI, Vestas was bringing in a partner with a huge balance sheet – MHI revenues were over Y3trn (US$0.5trn) in 2012 – that would put it on a par (at least financially) with the undisputed market leader in offshore wind power, Siemens. MHI also brought to the table expertise and industrial capacity in areas from shipbuilding to aerospace and space technology to power plants. The joint venture virtually guaranteed that the V164 would be built and deployed on a large scale, following two years in which Vestas' capacity to bring the machine to market had looked in doubt.

The deal was something of a turning point for the Danish company – the traditional front-runner in the wind industry – and a recognition of new realities. The company had been struggling to go it alone with the development of a new offshore mega-machine, while a bumpy financial performance had led to rumours that the company was about to be sold to a an industrial conglomerate, with Chinese companies being the most mentioned. A few weeks before the Mitsubishi announcement, Vestas had dismissed its controversial Chief Executive Ditlev Engel and replaced him with Anders Runevad, a Swedish telecommunications executive – the first time that Vestas had had a non-Danish

Figure 1.1 Vestas' giant V164 offshore turbine (Source: Vestas).

CEO. Runevad had come, in the words of Chairman Bert Nordberg – also a Swede and also a telecoms executive – to guarantee 'a future without surprises' for Vestas.

The deal with MHI was the latest chapter in a remarkable story that has seen a group of small, mainly Danish companies sell the wind-power concept to the world and turn it into a mainstream energy source and multi-billion-dollar industry.

From Tvindkraft to the California Wind Rush

In 1972, a group of radical teachers set up a collective and base on a plot of farmland called Tvind in the small Western Jutland town of Ulfborg, around 100km away from the then headquarters of a small agricultural equipment manufacturer called Vestas. Influenced by Maoism and the current debates around radical pedagogy, its charismatic and undisputed leader was Mogens Amdi Petersen. The group began to set up a number of Tvind schools across Denmark, which were supported by Denmark's liberal public subsidies.

In 1975, the Energy Crisis had led Denmark's government to consider a large-scale nuclear programme, and political support was growing for the idea, even though public opinion was largely hostile. Next door to Denmark, Sweden was about to commence production of power from its Barsebäck nuclear plant.

Figure 1.2 The Tvind group's charismatic leader, Mogens Amdi Petersen
(Source: Scanpix Denmark; photographer Aage Srensen).

Petersen and his group decided that the best way to mobilise public support against nuclear was not by protesting, but by showing that a practical alternative existed in Denmark's strong winds, and the Tvindkraft project was born. Tvindkraft foresaw the construction of a 2MW wind turbine, a size which was far bigger than anything that had been built before. The project mobilised volunteer labour and input from Danish universities.

Surprisingly the project was a big success. Up to 100,000 people visited Tvindkraft during the three-year building work, and when it was finished in 1978, the turbine worked. The design and construction of the turbine led to a number of technical advances which influenced the future development of the industry and had a big influence on some individuals who went on to become key figures in the modern wind industry.

Henrik Stiesdal, who went on to design Vestas first commercial turbine, built the first offshore wind project with fellow Danish company Bonus and became CTO of Siemens Wind, one of the world's biggest turbine OEMs (original equipment manufacturers); he was one of the people who visited Tvind. Stiesdal says,

> During Christmas 1976 my father and I therefore went to Tvind for the first time, and like everyone else were fascinated by this group of obvious amateurs, who from something that appeared to be absolutely square one, were building the world's biggest wind turbine.
>
> (Renewable Energy in Denmark 2000)

He said of his visit:

> The effect of the Tvind turbine as a source of inspiration cannot be over-
> stated. A large number of the pioneers became hooked, like me, on the
> possibilities and practical challenges of wind power when they visited Tvind.
> The almost nonchalant self-confidence with which the so-called 'Mill Team'
> built something no one had ever done before was very contagious.
>
> (Stiesdal, Personal communication)

Figure 1.3 Siemens Wind's CTO, Henrik Stiesdal, one of the pioneers of the modern
wind industry (Source: Siemens).

Stiesdal points out that some of the tools later commonly used in the wind industry, such as blade moulds and measuring tools, were developed at Tvind. It is significant that, in keeping with the Tvind group's political philosophy, the technology for the turbine and its blade was made publicly available – making it the first 'open-source' blade design long before the term came into use.

Among others who were influenced by Tvindkraft were the young engineers at the Risø DTU National Laboratory, which was originally set up by physicist Niels Bohr to study nuclear power and would go on to be a key institution in developing Danish wind-power technology and spreading it across the globe.

Before we move on, it is worth noting that the Tvind turbine is still producing power at the time of writing. The Barsebäck nuclear plant was shut down completely in 2005.

Vestas steps in

Henrik Stiesdal and blacksmith Karl Erik Jørgensen built a 10-metre diameter turbine in 1979 through a company called Herborg Vind Kraft or HVK. After a series of experiences with storms and lost blades, they, together with blade manufacturer Erik Grove Nielsen, developed a machine with pitchable blade tips to protect the turbine from reaching too high speeds. Meanwhile Danish agricultural machinery company, Vestas, had been experimenting with wind technology – mainly in secret – since 1971, with Birger Madsen in charge of engineering.

In 1979, HVK concluded that life as a stand-alone wind-turbine manufacturer was not a realistic option. At the same time Vestas' latest turbine idea – a vertical access 'whisk'-like machine, had resulted in failure, convincing Madsen that a new approach was needed. Stiesdal and his partner agreed to license their turbine design. This move resulted in the Vestas V10 and then the 55kW V15 turbine, which became its first truly commercial machine.

Planet Zond calling Denmark

Meanwhile, on the other side of the Atlantic, the threat to the US economy from the Arab oil embargoes had led President Jimmy Carter to create the US Department of Energy in 1977, and the government began to look at ways to develop wind and solar energy.

Federally funded R&D programmes aimed at getting big defence contractors to develop wind technology were largely unsuccessful, but federal and state regulatory policies and tax credits – such as the Public Utilities Regulatory Policies Act (PURPA) – which required utilities to take power from small non-utility generators, had a huge effect on the development of wind power.

California's Governor Jerry Brown implemented a state PURPA that ensured generous payment for power for wind developers, and developers also received a generous tax break through accelerated depreciation of their assets. The policies led to the California Wind Rush. Energy writer and analyst Daniel Yergin describes it in the following way: 'Committed wind advocates, serious developers, skilled engineers and practical visionaries were joined by flimflam promoters, tax shelter salesmen and quick-buck artists. Thus was the modern wind industry born' (Yergin 2011: 595).

Developers began setting up clusters of hundreds of machines in three giant wind-rich areas – the Altamont Pass, the Tehachapi Pass and the San Gorgonio pass – only to find that they were woefully ill-equipped to stand up to the wind they found there. Many of the turbines were destroyed soon after being installed, with blades flying off and towers collapsing; and most produced far less electricity than they were expected to. Many of the windfarm clusters became little more than eyesores.

One of the entrepreneurs most committed to wind power was James Dehlsen, who had founded Zond in 1980. Dehlsen spent New Year's Eve that year trying to install wind turbines in a blizzard in the Tehachapi pass in order to qualify for tax credits before they expired at midnight – something which as we shall see still sounds worrying familiar in the US wind sector.

Figure 1.4 US wind-power pioneer and Zond founder, Jim Dehlsen (Source: Clipper
　　　Windpower).

The outcome was disheartening. 'As soon as we started turning the turbines on they started disintegrating', says Dehlsen. 'The next day we picked up the pieces. We concluded we'd better get a better technology pretty damn quick' (Dehlsen, cited in Yergin 2011: 596).

Dehlsen travelled to Europe to look for turbines that were more robust. Finn Hansen, the son of Vestas' owner Peder Hansen, heard that Dehlsen was in the Netherlands and was about to buy Dutch turbines. He flew down in the company's twin-engine plane and persuaded the Americans to make the trip to Denmark, where they bought two Vestas turbines there and then. Zond followed up with an order of 155 Vestas V-15 turbines, 'much to the astonishment of Finn and his family', according to Dehlsen; while in 1982 it ordered 550. Vestas increased its employees from 200 to 870.

Danish companies supplied 90 per cent of the turbines during the Californian wind boom, with most of them supplied by Vestas, Bonus and Nordtank. The boom culminated with orders for 3,500 units in 1985. However disaster was not far away, as the California tax credits were about to expire. Zond had ordered 1,200 turbines to be delivered by 1 December 1985. According to Vestas, 'On the second shipment, disaster strikes: the shipping company goes bankrupt. Anchored outside Los Angeles, Vestas misses the deadline. When the turbines finally arrive, Zond refuses to accept them – and can't even pay for the turbines already delivered' (Vestas, history 1971–86: www.vestas.com/en/about/profile# !from-1971–1986).

Figure 1.5 The end of the first US wind boom left parts of California littered with dead wind turbines (Source: Eric Horst (Creative Commons)).

Dehlsen's description of the events of 1985 is rather different:

> We referred to that year as the D-Day of wind where the sheer logistics of multiple project locations, the number of turbines to be installed in the few remaining months of the year, with construction activities often hampered by blizzard conditions, had our team working around the clock. Thanks to my wonderfully supportive wife Deanna, who kept our field crews nourished with her chilli, we all survived the experience.
>
> (Dehlsen 2003)

In any case, by 1986 the wind rush was over, as tax credits were rolled back and oil prices fell, apparently ending the threat to US energy security. Zond was forced to cancel its planned initial public offering and negotiate with its creditor banks, going into 'full survival mode, liquidating equipment and painfully laying off a large part of our outstanding team' (Dehlsen 2003).

Back in Denmark, the government had made some – short lived – changes to the regulatory framework at the end of 1985 that were extremely unhelpful to the wind industry, requiring that people investing in cooperative wind farms had to be resident in the municipality where the wind farm was established. 'This was a kick in the back especially at Vestas where sales for Taendpipe, the biggest wind farm at that time were completed and owners were spread all over the country', says veteran wind-sector analyst, Per Krogsgaard. Vestas effectively went bankrupt and suspended its payments in October 1986. A new company, Vestas Wind Systems A/S was launched in 1987, with Johannes Poulsen as managing director.

Vestas bounces back

Interest in wind power was growing worldwide, and the company was soon manufacturing turbines for wind projects in India, backed by the Danish state-aid agency, Danida. In 1989, Vestas acquired government-controlled Danish Wind Technology, boosting its capacity and sales reach. The same year it created a German subsidiary and followed this with subsidiaries in Sweden and the US in 1992, while exports were also growing to the UK, Australia and New Zealand.

By the early 1990s, Vestas was selling turbines around the world, and was confident it could become the 'largest modern energy company in the world'. The technology on which the 'Danish concept' was built was evolving, with pitch control and lighter blades, and systems to ensure an even output of electricity to the grid. These innovations were encapsulated in its 1994 V44 turbine, which could produce 600kW of power. The company was pioneering the integration of large numbers of turbines into wind 'farms'.

In 1994, Vestas added more production capacity by buying Varde-based Volund Stalteknik. The same year it formed a joint venture with the Spanish aerospace and engineering company, Gamesa, giving it an entry into what would be one of its most important markets over the next 15 years.

In 1995, Vestas installed its first offshore turbines – the first offshore wind farm in Denmark had been installed four years earlier by fellow Danish turbine manu-facturer Bonus under Henrik Stiesdal's supervision.

In 1997, Vestas built what was then the world's largest commercial wind turb-ine, with a 1.65MW capacity, 55 times greater than the first wind turbine produced by the company in 1979. With 32-metre long blades, the turbine could produce enough power to supply around 1,000 households.

In 1998, with its turbines representing 22 per cent of the world market, the company carried out a stock listing on the Copenhagen Stock Exchange raising €175m (US$240m). The share offer was eight times over-subscribed.

Behind the growth of Vestas was steady political support at home. As early as in its second energy plan from 1981, the Danish government had published a target of 10 per cent renewable electricity, to be met by a combination of 60,000 wind turbines and 5,000 decentralised biogas plants. Denmark had cancelled its plans for nuclear power in 1985, a year before the Chernobyl nuclear disaster. Inspired by the 1987 'Brundtland Report' – Our Common Future – Denmark began to set targets for wind-power growth as part of its overall plans to limit carbon emissions. The third Danish energy plan, Energi 2000, presented in 1990, called for 1,500MW of wind power to make up 10 per cent of electricity consum-ption by 2005 – a share that was reached in the spring of 1999. The fourth energy plan from 1996 added a long-term 2030 target of 5.5GW, of which 4GW should be installed offshore, for wind power to meet 50 per cent of Denmark's power demand.

In the event, championed by Energy and Environment Minister Svend Auken and backed by broad political agreements on Danish energy policy, the wind industry was able to exceed these targets. Capacity grew from 400MW to 1.4GW between 1995 and 2000, when it already met 12.5 per cent of the country's power consumption. A new right-wing minority government, led by Prime Minister

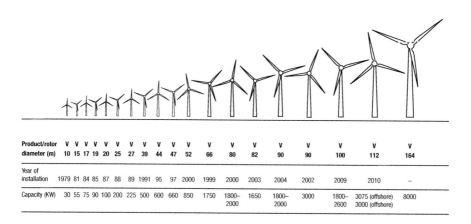

Product/rotor diameter (m)	V 10	V 15	V 17	V 19	V 20	V 25	V 27	V 39	V 44	V 47	V 52	V 66	V 80	V 82	V 90	V 90	V 100	V 112	V 164
Year of installation	1979	81	84	85	87	88	89	1991	95	97	2000	1999	2000	2003	2004	2002	2009	2010	–
Capacity (KW)	30	55	75	90	100	200	225	500	600	660	850	1750	1800–2000	1650	1800–2000	3000	1800–2600	3075 (offshore) 3000 (offshore)	8000

Figure 1.6 Growth in size of Vestas' turbine models, 1979–2014 (Source: Vestas).

Anders Fogh Rasmussen, took over in the autumn of 2001, ending the long-held tradition of broad political agreements on Danish energy policies. With the support of the right-wing Danish People's Party, the coalition government had a narrow majority, which it used to dismantle the progress made under Svend Auken and former Prime Minister Poul Nyrup Rasmussen. This led to five years of relative stagnation in Denmark from 2003 to 2008.

The leader of the Conservative Party – former policeman Bendt Bendtsen, who is currently a member of the European Parliament – was given the energy portfolio in the new coalition government. In February 2002, he told Reuters that wind energy was too expensive and that other countries should be cautious about wind energy's impact on their competitiveness: 'I'm of the opinion that Denmark shouldn't continue to subsidise installation of new wind turbines after 2003.' His opinions went worldwide and the leading Danish companies found it difficult to understand why the new minister found it necessary to hurt exports, in addition to shattering the Danish home market.

'It was export masochism', says Christian Kjaer who was a policy advisor to the Danish Wind Energy Association at the time. 'It was also bizarre that the minister responsible for trade and industry warned about competitiveness at a time when Denmark, according to the World Economic Council, was the world's fourth most competitive nation' (Kjaer, personal communication).

A meeting held later in 2002 between Minister Bendtsen and frustrated CEOs of the Danish wind-energy companies was a farce. The new minister told the CEOs that it was not he but his larger coalition partner – The Liberal Party – that was behind the changes in policy. The CEOs left the meeting in despair, waiting for better days. They had to wait until 2007, when Connie Hedegaard was appointed Minister for Climate and Energy. She even managed to convince Prime Minister Anders Fogh Rasmussen that the coalition's policies had been a mistake. In 2008, referring to his own party's energy policies, he admitted it had been dragging its feet. 'Let me do something that I know many would like and have been waiting for me to say: that was probably wrong', he told his party's national congress in 2008 – a year before world leaders were to arrive in his capital Copenhagen to reach a global agreement on emissions reductions.

Meanwhile, momentum around limiting carbon emissions had been growing steadily outside Denmark, leading to the passing of a growing body of legislation to extend support for renewables between 1990 and the present day. Following the 'discovery' of the climate issue in the second half of the 1980s, the German Enquete commission (The Enquete Commission on Preventive Measures to Protect the Atmosphere) was established in 1987, and came out with two seminal reports, one in 1989 and one in 1990, which was the beginning of the 'Energiewende' (energy transition) in Germany. It was also was one of the first major moves in the war between the pro- and anti-nuclear forces in Germany, the latter of which had been strengthened due to Three Mile Island and more importantly, the Chernobyl disaster.

Following these reports, the first German renewables support legislation came in, the 1991 Electricity Feed in Act, the forerunner of the EEG. GWEC

General Secretary, Steve Sawyer, calls the Feed in Act 'probably without much doubt the single most important piece of national legislation giving rise to the modern renewables industry, particularly wind' (Sawyer, personal communication).

Spain passed a landmark electricity law in 1997, setting the stage for that country's subsequent boom in wind power. But the key moment in terms of scaling up the level of ambition for the renewable sector was the original EU renewable electricity directive of 2001, setting targets for all the then 15 member states of the European Union. This directive had a target of increasing the share of the EU's electricity from renewables from 14 per cent to 22 per cent by 2010. Another important development was the passing and implementation of China's Renewable Energy Law in 2005–06. 'While the drivers were initially energy security, they were enhanced throughout the 90s and the first decade of the twenty-first century by climate concerns, and the jobs and industrial development and local economic benefits drivers came along with the scaling up', says Sawyer (personal communication).

Even more compelling was the '20:20:20' EU Climate and Energy package legislation which was agreed in December 2008 and finally entered into force on 25 June 2009. 'Today tomorrow changed! The European Parliament and the Council have agreed the world's most important energy law', said then then EWEA Chief Executive, Christian Kjaer, on the day of the agreement.

The agreement had been preceded by a rigorous battle at the highest levels of European politics. At a meeting between European Commission President José Manuel Barroso and then president of the European Wind Energy Association (EWEA), Arthouros Zervos. Zervos had asked for sectoral renewable targets for electricity, heating and transport, because an overall binding renewable energy target seemed impossible to get through the Council of Ministers. The Commission President told Zervos that he would be able to deliver a binding overall 20 per cent target from the Member States, if the industry compromised on its demand for sectoral targets. Zervos replied that this would be acceptable for the industry, but nobody believed it possible to have all 27 governments agree on a binding target.

In January 2007, only three Member States – Germany, Sweden and Denmark – supported the idea of having a binding target for renewable energy. EU energy ministers meeting on 15 February failed to reach an agreement and left the decision to the 27 Heads of State, who met three weeks later. On 9 March, led by the German Presidency of the EU, they unanimously agreed 20 per cent binding targets for renewable energy and carbon reductions and an indicative target of 20 per cent energy efficiency, all to be reached in 2020.

'The unanimous adoption by 27 European Heads of State could not have happened at any other point in time', Christian Kjaer recalls today. 'Oil prices had just risen above $100 per barrel for the first time, Russia and Gazprom had cut off gas supplies to Europe a few months before, the economies were booming and Germany held the Presidency of the European Union' (Kjaer, personal communication).

The agreement had three main elements: a carbon reduction target, a renewables target and an efficiency target. 'In terms of legislative quality, it was the good, the bad and the ugly', Kjaer says.

The renewables legislation was excellent; the greenhouse gas reduction target of 20 per cent was undermined by the fact that external carbon reduction credits could be used, meaning that effective reduction within the EU was below 10 per cent, compared to the IPCC's demand for 25–40 per cent. But the really ugly one was the 20 per cent energy efficiency target, which was merely indicative and, thus, could not be enforced in Court (Kjaer, personal communication). Ten months later, the European Commission tabled its legislative proposals in January 2008.

During the ten months following the Heads of State's political agreement on targets, an intense battle was being fought between the European renewable energy industries, united under its umbrella organisation EREC (European Renewable Energy Council), and large industrial players within and outside the energy sector, gathered under the umbrellas of Eurelectric and Business Europe, who lobbied for the Commission to come out with a weak – or even better an outright unworkable – proposal from the Commission. The battle was also going on inside the Commission, between civil servants of opposing views.

'We were losing the battle of the Commission proposal for 9 months', Kjaer says.

> It was only the direct personal intervention by then Energy Commissioner Andris Piebalgs during the last month of drafting, that turned the tide, and a good proposal was tabled in January 2008. Claude Turmes, one of the European Parliament's best legislators, had been appointed Rapporteur on the Renewables Directive and he sealed a great final outcome of the negotiations with the Council in December 2008.
>
> (Kjaer, personal communication)

The 'Danish concept' goes global

By 2000, demand for wind power around the world had taken off and, led by Vestas, exports went from strength to strength. The world market had grown to almost 4GW of annual installations, of which Vestas was supplying almost a third. The 'Danish Concept' had gone global.

As different countries began to develop their wind capacity, a number of turbine companies emerged to compete with Vestas. In Spain, Gamesa Eólica was formed in 1994 to supply wind turbines and develop wind farms from an engineering group that had mainly concentrated on aerospace. Vestas was involved, as a partner, with a 49 per cent shareholding and technology licensing agreements. The licensing agreement between Vestas and Gamesa limited the Spanish company to selling its turbines in Spain, Latin America and North Africa, but Gamesa's management had global ambitions, and there were conflicts over development activities carried out by Gamesa in places like Italy and Greece.

In 2001, Vestas sold its stake in Gamesa for €287m (US$395m). Vestas' CEO Johannes Poulsen said at the time:

> Differences in strategy between Vestas and Gamesa have led to an increasing number of strategic conflicts in the marketplace. We have therefore been looking for ways and means to avoid this and our sale of Gamesa Eólica shares seems to be the better option for all parties involved.
>
> (*Windpower Monthly* 2001)

Under the terms of the agreement, Gamesa was able to continue using Vestas' technology, including the new V80-2MW turbine, while Vestas gained access to the fast-growing Spanish market. 'We are creating ourselves a new competitor for whom we have the highest respect as we know his industrial capability as well as his basic technology', said Poulsen. 'But we prefer this over a situation that creates uncertainty about Vestas strategy and we believe the market for wind power gives room for both of us in the future' (*Windpower Monthly* 2001).

Gamesa went on to sell its aerospace division in 2006, and in the same year was in second place in the global wind-turbine supplier rankings with over 10GW installed and a market share of over 15 per cent.

In Germany, engineer Aloys Wobben had established Enercon in 1984 and built the company's first turbine, the 55kW E-15/16. In 1993, Enercon produced its E-40/500kW series using a highly original direct-drive configuration with an annular generator that eliminated the need for a gearbox, which had historically been one of the most common areas of mechanical failure. Enercon turbines proved to be extremely successful among the small developers who are the driving force in Germany's wind market, due to its reputation for reliability and its highly popular PartnerKonzept service model. Enercon's attempts to expand internationally after 1996 were less consistent than those of Vestas and Gamesa, and Wobben found himself embroiled in a number of patent disputes with rivals such as Vestas and Kenetech/Zond that hampered operations, particularly in the US. However, the company built up operations in Sweden, Brazil, Canada, Portugal, France and Austria. It has consistently been in fourth or fifth place in the global supplier rankings, helped by its dominant position in Germany.

Wind-sector consultant and analyst Aris Karcanias notes of Enercon: 'They have managed to consistently maintain a leading position amongst the top-five turbine OEMs despite not competing in the world's two largest markets, the US and China, and actively deciding not to develop turbines for the offshore market' (Karcanias, personal communication). He adds that Enercon is considered by many to be the 'Apple' of the wind industry because of its pioneering of a direct-drive wind turbine without the use of permanent magnets (which proved to be a pinch point for other OEMs in later years), as well as concrete towers.

As we will see, as wind markets developed outside Europe, they developed their own national champions, as in the case of India's Suzlon, which was founded in 1995 and expanded rapidly thereafter, and China's Goldwind, founded in 1998.

Ditlev Engel and 'The Will to Win'

In early 2004, Vestas took over NEG Micon, the second-largest Danish wind company. The merger allowed Vestas to achieve a global market share of 34 per cent and revenues for the year of €2.56bn (US$3.53bn). Commentators saw the mergers as a defensive move by Vestas that would prevent NEG Micon from being taken over by a company from outside Denmark, including Gamesa, which was making increasingly aggressive moves in the market after its split from Vestas a few years earlier. Critics said that Vestas had overpaid for the poorly performing NEG Micon, calling the deal a rescue rather than a merger. The merger deal had been under discussion for over a year, but NEG Micon reportedly 'surrendered' to a takeover by Vestas after it was forced to issue a profit warning in November 2003 (*Windpower Monthly* 2004). Managers of the combined company said they were targeting savings of €67m (US$92m) a year through the reduction of overlapping sales and production facilities as well as better deals with suppliers of components and raw materials.

In October the same year, Vestas hired Ditlev Engel as its new President and CEO to replace Svend Sigaard. Engel had worked his way through the ranks – and through business school – at Danish paint manufacturer Hempel, and become the company's CEO in 2000 at the age of 35. Without any knowledge of wind power and coming from outside the engineering fraternity that dominates the sector, Engel says he spent the two months before he took up his new post in May 2005 travelling the world and getting to know the business.

On 26 May, he presented the company's first-quarter results along with a new strategic plan with the Nietzschean name 'The Will to Win'. Key to the vision was a plan that would supposedly take wind power from being a still relatively marginal source of energy to one that was on a par with oil and gas.

'Many people still regard wind power and thereby Vestas as a "romantic flirt" with alternative energy sources. It is not. Vestas and wind power is a real and very competitive alternative to oil and gas', said Engel.

Engel's strategy promised, among other things, that Vestas would achieve:

- earnings before interest and tax (EBIT) of at least 10 per cent by 2008 from an expected 4 per cent in 2005
- a global market share of 35 per cent
- a reduction in production costs including a staff reduction.

Along with the new targets and vision, Engel created three new business units to add to Vestas' expanding corporate structure – they were Technology, People and Culture and Offshore, which were added to Nacelles, Blades, Northern Europe, Towers, Mediterranean, Asia-Pacific and Americas – while centralising decision making in an Executive Board of Management that had two members, himself and Executive Vice-President and CFO, Henrik Nørremark. As we shall see, the relationship with Nørremark, who unlike Engel was a Vestas veteran, would play a fundamental role in Engel's career in Vestas.

Figure 1.7 Ditlev Engel, who presided over wholesale expansion at Vestas after taking over as CEO in 2005 (Source: Vestas).

In language that was part business management speak and partly a throwback to the heroic discourse surrounding the Tvindkraft project, employees were urged to increase their efforts, with Engel saying 'This work will no doubt be exciting and very hard. At the same time, it will require the will to change for all of us.' At the same time a sculpture called 'The Willpower' was commissioned from artist Jørgen Pedersen and copies were installed outside Vestas' offices around the world (see Figure 1.8). Corporate literature described the significance of the sculpture in the following terms:

> Vestas was founded to take on one of the biggest challenges facing the earth. We have never been, and will never be, characterised by losing faith and commitment when met with resistance and inertia. On the contrary, Vestas and our employees have shown persistence in breaking down misconceptions towards wind energy and supporting the growth of the industry. We call this our Willpower. It is expressed in the sculpture entitled 'The Willpower', which is placed at a number of the Group's locations. Reaching for the sky, it symbolises the willpower and passion possessed by the employees.

Figure 1.8 'The Willpower' sculptures were installed outside Vestas' main offices under Engel (Source: Vestas).

For the next eight years, Engel would become the best-known face of the wind industry. As well as becoming a celebrity in Denmark, he appeared in big financial forums such as the World Economic Forum in Davos, took part in working groups run by the United Nations and the IEA, and was a regular on business news channels like CNBC, CNN and Bloomberg TV. There is no doubt that he made a big contribution to raising the profile of the wind industry and getting it taken seriously in the wider world of politics and business as a real alternative to fossil-fuel power generation. Engel kept a frenetic timetable, flying by private jet from country to country. Quarterly financial results were held alternately in London and New York, for instance.

Engel's leadership was both dynamic and charismatic. However, the consensus is that it was not successful in terms of reaching the objectives that he had set out in 'The Will to Win'. Despite talking initially of a reduction in staffing, Engel presided over a more than doubling of Vestas' head count, which reached a high point of 23,252 in 2010. Vestas opened its first factory in China in 2006, and its US blade, tower and nacelle plants in 2008–10. It added several more factories in European locations including the UK, Italy, Spain and Germany, along with a network of R&D centres.

Revenues and annual orders soared as the company tried to grow at full speed to accompany the growth of the wind market. Engel did reach the target of a 10 per cent EBIT margin in 2008, the year when Vestas' profits hit their peak and wind was effectively a 'seller's market', with demand for turbines outstripping global manufacturing capacity. But, as the rate of market growth – outside China – began to slow, capacity outstripped demand and prices and margins began to fall again.

Meanwhile Engel's ambition of maintaining a market share of 35 per cent proved impossible as competition intensified. Vestas found itself being squeezed, particularly in China by fast-expanding local producers, and in the US, where there was intense competition from both fellow European producers and from GE. Vestas' ambitions to maintain its market share were an important factor behind a deterioration in margins on turbine sales, as the company entered into contracts at unfavourable prices between 2009 and 2011.

On 27 October 2009, on the eve of the Copenhagen climate talks, Engel announced new, even more ambitious targets for Vestas; the so called 'Triple 15'. These foresaw achieving an EBIT margin of 15 per cent and revenues of €15bn (US$21bn) no later than 2015. As the company literature noted, this translated 'into an average annual growth of at least 15 per cent and a substantial improvement in the EBIT margin'.

Table 1.1 Vestas' key financial indicators

Year	Revenues (€ m)	Ebit margin (%)	Net profit/(loss) (€ m)
2004	2,363	(2.1)	(61)
2005	3,583	(3.2)	(192)
2006	3,854	4.9	113
2007	4,861	5.3	104
2008	5,904	10.4	470
2009	5,709	4.9	125
2010	6,920	4.5	156
2011	5,836	(1.0)	(166)
2012	7,216	(9.7)	(963)

Source: Vestas.
Notes: Figures were revised after Vestas changed its accounting procedures in 2010.

Table 1.2 Vestas' turbine deliveries

Year	Deliveries (MW)
2004	2,784
2005	3,185
2006	4,239
2007	4,502
2008	6,160
2009	6,131
2010	4,057
2011	5,054
2012	6,171

Source: Vestas.

Analyst Robert Clover says that 'Vestas' triple 15 goals looked ambitious when they were announced, as there were already signs of market weakness starting to appear then' (Clover, personal communication). He notes that equity markets never seemed to believe the targets would be met, with analyst consensus forecasts 'way below' Vestas' margin and sales targets. 'Had the market believed triple 15, then the stock would have been valued in excess of DKK 1,000' (Clover, personal communication).

Just a few months later, with the failure of the Copenhagen talks and demand growth beginning to slow as the financial crisis hit home, it started to become clear that Triple 15 had been a mistake, and in 2010 Vestas saw its first significant drop in deliveries since the aftermath of the California Wind Rush.

The wholesale expansion that Vestas carried out under Ditlev Engel was understandable, given the speed of growth of the global wind market in 2005–09, but it had left the company overexposed. Most analysts are also highly critical of Vestas' control over costs during this period. As we noticed, Engel had added more business divisions to the company's already complex structure, adding new

Table 1.3 Vestas' share of the global market (2005–12)

Year	Market share (%)
2004	34.0
2005	27.9
2006	28.2
2007	22.6
2008	19.8
2009	12.5
2010	14.9
2011	12.9
2012	14.0

Source: BTM-Navigant.

layers of middle management, with many of the divisions acting as complete companies in their own right. And, as would become apparent in 2011 and 2012, there were serious lapses in the company's cost calculations and the execution of its new technology programmes.

All this left Vestas vulnerable, and paved the way for a series of profit warnings in 2011 and early 2012 that shocked investors, and a rapid deterioration in the company's finances. By 2012, the EBIT targets of 10 per cent and then 15 per cent had turned into a negative margin of 9.7 per cent and a loss of almost €1bn (US$1.4bn). The main factor behind Vestas' growing problems was increased competition. With the wind market gaining momentum throughout the decade and growing at a rate that surprised even its most enthusiastic advocates, and returns being high, Vestas and the other pioneering companies were not going to have the wind-power market to themselves for long, as some of the world's biggest industrial companies made their moves.

2 Big industry moves in

GE sweeps up Zond

As we have seen, James Dehlsen was left licking his wounds at the end of the California Wind Rush. The US wind industry had gone into a deep decline, with most of the companies going bust.

Dehlsen had the advantage that Zond had shares in all the projects he had developed, and he spent the next years re-engineering the California projects that had already been built; and says he succeeded in raising output by 22 per cent between 1986 and 1989. He also persuaded utility Florida Power and Light (FPL) to invest in Zond's remaining projects in the Tehachapi Valley, leading to the building of the 77MW Sky River facility, which at the time was the largest single project in the US.

Dehlsen says that surmounting major obstacles – including the demand from utility Southern California Edison that the developers provide 75 miles of 220kV transmission lines, adding US$30 million to the project cost – and completing Sky River meant that 'Zond's survival was now assured, and we could finally make good to the banks and our other generously supportive creditors who had grown to share our vision' (Dehlsen 2003).

The 1991 Gulf War and the Energy Policy Act of 1992 had seen the reintroduction of tax credits for wind power, but with the important difference that the new credits rewarded the actual production of electricity from wind projects and not just investment in building new turbines. As the 1990s went on, different states began to implement renewable portfolio standards, and the US wind industry began to pick up speed again.

The question for Zond was now how to build its own turbine to lower the cost of energy and allow it to capture the manufacturer's profit margin, as Dehlsen saw it. The company received support from the National Renewable Energy Laboratory (NREL) and the DOE in 1993 for work on Zond's 550kW Z40 machine, which was completed in 1995. Zond then acquired patent rights from the US's biggest and best-known turbine manufacturer, Kennetech, when it went bankrupt in 1996, as Dehlsen pursued variable-speed technology. Zond produced a 750kW turbine and then started work on a 1.5MW turbine.

Dehlsen says he approached the then growing energy giant Enron in 'anticipation of the company's capital needs for project development and
manufacturing ramp-up' (Dehlsen 2003) and sold Zond for US$100m. Enron
officials say that they had decided to make a play on wind energy, and 'brought
Zond back from the brink' (former Enron official Robert Kelly, cited in Yergin
2011: 601). In any case, on Dehlsen's urging, Enron then acquired the bankrupt
Germany turbine manufacturer Tacke, and by combining technology from the
two companies, Zond produced the TZ 1.5MW turbine. Although Enron Wind's
sales soared between 1997 and 2001, in autumn 2001 Enron collapsed spectacularly into bankruptcy and scandal.

Industrial giant General Electric had been considering getting into the wind
business for several years. Two key executives – Mark Little and Jim Lyons –
unsuccessfully pitched the idea of a major move into the sector in the late 1990s
under former CEO Jack Welch. 'When Jack was CEO we did not get a warm
reception', says Little. 'He did not see wind as a serious business' (Magee 2009:
50).

Just ahead of the Enron bankruptcy, however, a dynamic new CEO, Jeff
Immelt, had taken over the leadership of GE. While cautious about the high cost
of entry into the wind business, Immelt had already understood the long-term
strategic potential of the wind industry.

In 2002, GE stepped in and bought Enron's wind business for US$328m,
creating a new GE Wind Division. In its second-quarter earnings report of 2002,
GE said that the deal 'established GE Wind Energy as a leading presence in the
renewable-energy industry, which is growing at nearly 20% a year'. Note though
that GE subsequently went to a court to ask for a refund of 50 per cent of the
purchase price (see Mumma 2002).

While the technological elements were there, GE had to invest heavily and
bring a lot of engineering expertise to bear to create a 1.5MW turbine that would
meet its strict standards. 'The industry was fundamentally broken', said the
former head of GE Wind, Vic Abate. 'We need to go through and rigorously re-
engineer to make it a mainstream technology.' Immelt goes even further.
'[Enron's wind business] was a very broken model that took us three years to fix',
he says (Magee 2009: 54). When the task was completed, however, GE had a
turbine that would prove to be one of the best-selling models in the world, one
which would make a serious dent in the sales of the Danish incumbents.

By 2005, the US wind industry was picking up speed fast, with the annual
installed capacity growing at an average rate of 40 per cent between 2005 and
2009. In 2009, GE's 1.5MW machine made up half of the entire wind-turbine
fleet in the US, with 10,000 units installed. It followed the 1.5MW machine
with the larger GE2.5XL, which was installed in the giant 845MW Shepherds
Flat wind farm in 2011.

GE also made a foray into offshore wind, acquiring Norwegian turbine developer ScanWind in 2009 and installed a 4.1MW offshore prototype in
Gothenburg, Sweden at the end of 2011. The venture was quietly shelved less
than two years later, but as we shall see, GE was to later make a far more

Figure 2.1 GE turbines at a site in the US (Source: GE).

aggressive move into offshore, acquiring France's Alstom Wind and its 6MW
Haliade turbine in 2015.

GE's wind division has consistently been in the top four turbine manufacturers, helped by its global financing and service capabilities. In 2012, after a
record year that saw around 13.2GW installed in the US, analysts at BTM-
Consult and IHS-Emerging Energy Research estimated that GE had knocked
Vestas off its number one spot, with a global market share of 15.5 per cent.

On the other hand, it is worth noting that GE's ability to win market share
outside the US has been relatively limited, although it has done well in individual markets such as Turkey and Brazil. In 2012, 75 per cent of the turbines GE
delivered globally were in the US, where it commanded a 38 per cent share of
the market. By contrast, in Europe its market share sat at about 6 per cent,
trailing its global rivals. Three of them – Vestas, Siemens and Gamesa – all held
far higher market shares in the US last year than GE did in Europe.

In 2013, Anne McEntee, a veteran of GE's oil and gas business, replaced
Abate as the head of GE's renewables business. GE also appointed Cliff Harris as
its new general manager for renewables across Europe, the Middle East and
Africa with the task of increasing the company's market share sharply in those
markets. Though GE has local manufacturing operations in Salzbergen, Germany, growing market share organically in European markets is a tough ask in a
market dominated by strong local competitors.

In 2015, with wind-turbine markets once again buoyant, GE launched
another major acquisition, buying Alstom's energy business for around $10.6bn.

The deal took months of protracted negotiations, having sparked a rival bid from German manufacturing giant Siemens (see Chapter 9).

The acquisition bought GE Alstom's sizable onshore wind business – the French company was particularly strong in Brazil where it had become the leading manufacturer – and also propelled GE back into the offshore business. However, it was not alone in looking to grow through acquisition, and we will look at GE's position post-merger later in this book.

Siemens Wind goes for scale

We have already met Henrik Stiesdal at the Tvindkraft project and in the earliest development of the 'Danish concept' and Vestas' first viable turbine. After working as a consultant at Vestas and then taking time off to complete his studies, Stiesdal joined Danish wind-turbine manufacturer Bonus Energy in 1987, first as a development specialist and from 1988 technical manager, becoming Chief Technology Officer in 2004 and playing a key role in building Siemens Wind into a global turbine manufacturing giant, before retiring from the company in 2014.

In 1990–91, Stiesdal was responsible for building the world's first offshore wind farm at Vindeby, Denmark, with 11 specially adapted 450kW turbines. In 1998, he designed Bonus's first variable-speed turbine and introduced a series of innovations in blade manufacturing and control.

In 2004, the Bonus management sold the company to German industrial giant Siemens for an undisclosed amount that was estimated at anything between US$240 and US$400m, with Stiesdal staying on as CTO. The company maintained its manufacturing operations in Brande, Denmark. Analysts said that Siemens had made the move after closely watching GE, and in the fear that it would lose out to its US rival in a strategic growth area. Like GE, Siemens was looking to compensate for an expected fall-off in orders for conventional power-generation equipment, like gas and steam turbine plants, with rising sales in wind equipment.

Andreas Nauen, then vice-president of communications and strategy at Siemens Power Generation (and later to lead first Siemens Wind Energy and then REpower), said: 'We wanted to find a company that has the same business approach as us and that would complement our business.' Nauen also noted that 'Bonus has a superb reputation, is very risk aware and its technology is a good fit' (quoted in Hoel 2004). Siemens was particularly interested in building larger turbine models, and Bonus was already testing a 3.6MW turbine design for offshore, Nauen pointed out.

Following the Bonus acquisition, Siemens betted big on the still small but potentially massive offshore wind sector (see Chapter 5). It was able to avoid some of the highly costly failures offshore that Vestas suffered, and its 2.3MW and then 3.6MW offshore turbines became highly regarded. Having the solid financial backing of a company of Siemens' size was also a major advantage; by 2009, the company had a market share of nearly 75 per cent of the installed offshore market, rising to nearly 85 per cent by 2012.

Bonus had been a much smaller company than Vestas or NEG Micon, but also a more profitable one. Onshore, Siemens grew more slowly, with a step-by-step ramp-up of its production facilities and a more cautious approach than Vestas when deciding to set up facilities in new markets. However, by 2012 it had jumped into third place in the global rankings, with 9.5 per cent of total installations, according to BTM-Consult.

Wind power grew from 0.5 per cent to 5 per cent of Siemens' turnover between 2004 and 2011, with Siemens Wind increasing its employees from 800 to the current level of around 8,000. Most of its production is still carried out in Brande, but it also set up facilities in the US in 2009 and 2010, as well as blade and nacelle factories in China and a blade-manufacturing facility in Canada. As we shall see, it also set up a major joint venture with Shanghai Electric to facilitate its entry into China's market and made a final investment decision to set up a major new factory to produce offshore windturbines in the UK in March 2014.

In mid-2008, Siemens began testing direct-drive turbines and in 2009, it installed a 3MW prototype near Brande. This was followed by a 2.3MW version and a giant 6MW offshore direct-drive machine, with the first offshore prototypes deployed in DONG Energy's Gunfleet Sands project in the UK in 2013.

In 2011, the Renewables Division of Siemens Energy was split and Siemens Wind Power became a separate division with headquarters in Hamburg, Germany. The majority of offshore and a significant part of onshore competences remain in Denmark and Stiesdal remained CTO, giving the company remarkable continuity in terms of technological development and understanding of the wind sector's evolution. Of the current players, Siemens is a good bet to be one of the top three, or perhaps even the top wind-turbine company, in the coming years.

Alstom and Ecotècnia

French engineering giant Alstom has long been a heavyweight in hydroelectric power-generation equipment, as well as transport, power transmission and other areas. In mid-2007, it acquired Spanish turbine manufacturer Ecotècnia for €350m (US$480m).

Founded in 1981 and with its headquarters in Barcelona, Ecotècnia was one of the leading Spanish turbine manufacturers along with Gamesa and Acciona Windpower and it had a global market share of around 2 per cent, with most of its installations in Spain. Its turbines were based on a robust design meant for the turbulent winds of mountainous Spanish areas, which isolated the gearbox from the main drive, reducing non-torque gearbox loads, and a modular construction that was intended to facilitate construction in hard to access areas. These designs became the basis for Alstom's turbine portfolio, and the company scaled up the size of its machines and blades, producing the first ECO100 3MW turbine in 2009. The company was renamed Alstom Wind in 2010. The same year, Alstom began constructing a turbine nacelle factory in Amarillo, Texas, to add to its assembly plants in Somozas and Buñuel in Spain.

Figure 2.2 Siemens' SWT3.6 has been the best-selling offshore turbine in the world
(Source: Siemens).

As well as winning some major onshore orders in Europe, such as the second phase of Iberdrola's giant Whitelee wind farm in Scotland, Alstom Wind's main success has been in Brazil, where it has been the fastest growing turbine company in recent years (see Chapter 4).

Alstom has also moved into offshore, with its 6MW direct-drive Haliade machine, which it began testing in early 2012. Alstom has yet to register in the ranking of the top ten turbine manufacturers and the company seems to be taking a cautious approach to entering new markets. Officials maintained throughout 2012 and 2013 that the company had the ambition to be one of the top three or four players in the world and its experience in Brazil showed a willingness to make aggressive moves when it saw that the opportunity is there. However, Alstom's management subsequently decided that the demands of becoming a world leader in energy were too much for the company's balance sheet, and as we have seen, the French company decided to sell to GE in 2015.

Offshore wind, with its very big demands in terms of capital expenditure and warranty risk has shown itself to be an opportunity only for companies with strong nerves and balance sheets. As well as Alstom, offshore wind attracted the entry of French nuclear energy company Areva, and (for a while at least) the South Korean chaebol Samsung and Hyundai. Japan's Mitsubishi (now as part of a joint venture with Vestas) and Hitachi have also pushed into offshore wind, along with a number of Chinese companies.

Figure 2.3 Alstom's 6MW Haliade offshore turbine being erected (Source: Alstom).

Life gets harder for the 'pure players'

The outcome of the entry into the market of big industrial companies like GE, Siemens and Alstom has been to make life harder for the 'pure play' wind-turbine producers such as Vestas, Gamesa and Suzlon. The diversified groups can often count on ongoing relationships with big utilities, as well as the ability to offer finance through other group divisions, but the key factor is bigger balance sheets to cope with any claims on warranties due to turbine failure, particularly in the offshore arena. Arguably, the diversified companies are able to cope better when there are lean years in the wind sector, because they are often naturally 'hedged' through their involvement in other energy-generating technology.

In short, a couple of bad years in the wind sector are not going to have the same effect on Siemens or GE as they will on Vestas or Gamesa, although it is important not to exaggerate this point, as parent companies have shown that they will not tolerate consistent losses from their renewable energy divisions – as Siemens proved when it shut down its loss-making solar business in 2013. Some of the diversified companies can also bring considerable R&D resources into play as well, and leverage their industrial set-ups. GE is the most obvious example, with its research labs working on everything from fabric blades to better storage and 'intelligent wind farms', while it has also been able to divert staff and facilities from making gas turbines to making wind turbines during boom years, before re-assigning them to their original duties.

Wind development scales up as utilities come in

One factor that was important in attracting large-scale industrial companies to the wind market was the increasing scale of wind developers, as a series of large-scale utilities and power companies entered the sector. The increasing size of projects and investors meant that there were increasing opportunities for large-scale contract wins rather than large numbers of small turbine sales. The entry of utilities into the business also meant that the likes of GE, Siemens and Alstom were talking to the same companies that they were selling to in their other power businesses.

The rise of very large-scale wind developers was led initially by a small group of companies in Europe and the US. The most notable is Spanish utility Iberdrola, which formed its Iberenova business in 2001 and built it – through a mixture of organic growth and acquisitions – into the largest wind-power company in the world with a capacity of over 7GW by 2007, and a growing presence in the Americas and Europe. That same year, Iberdrola floated Renovables on the Madrid stock exchange, raising around €4.5bn (US$6.2bn).

In the US, utility Florida Power and Light was an early mover into wind power, despite there being no wind projects in Florida. The company had invested in wind projects as part of a diversification programme in the 1980s, and found itself owning a number of wind farms when a number of these projects went bankrupt. The company, now known as NextEra Energy Resources, stuck with the wind business throughout the 1990s becoming the biggest developer in the US – accounting for around 20 per cent of total wind generation at one point – and one of the biggest in the world, with 10.21GW of installed wind capacity as of 31 December 2013.

In Europe, Iberdrola was among a group of European utilities that built their renewable energy arms into dynamic operations and in many cases floated renewables subsidiaries on the stock market. These include Portugal's EDP (EDP Renováveis); Italy's Enel (Enel Green Power); France's EDF (EDF Energies Nouvelles) and GDF Suez; Germany's E.ON; as well as the Spanish infrastructure company Acciona, which built Acciona Energia into a massive pure-play renewables power producer. Also notable are Denmark's DONG Energy, which has become the world's first utility whose development is centred on offshore wind; and Warren Buffet's MidAmerican Energy Holdings, which has become a major US wind player in recent years.

These companies have been joined in recent years by a group of Chinese generators, which include China Lonyuang Power Group, China Huaneng Group, Datang, China Huadian Corporation, China Guodian Corporation and China General Nuclear Wind Power. These Chinese companies have pushed into the ranking of the world's top developers, occupying four out of the top ten positions and ten out of the top twenty-five spots.

Figure 2.4 Iberdrola's CEO, Ignacio Sanchez Galán, transformed the Spanish utility into the biggest wind-power operator in the world (Source: Iberdrola).

Table 2.1 Iberdrola's installed renewables capacity

Year	Capacity (GW)
2007	7,098
2008	9,302
2009	10,752
2010	12,532
2011	13,690
2012	14,034
2013	14,247

In 2012, the largest global wind owners accounted for 117.4GW, according to analyst IHS-EER, or fully 43 per cent of cumulative global installed capacity. The effect that the development of large-scale wind-based utilities had on the industry cannot be underestimated. Wind projects had been increasing progressively in scale throughout the 1990s. Utilities, with their ability to finance projects on their own balance sheets, could increase the size of wind farms to an extent few had hitherto imagined.

As well as being able to finance wind farms, the big power producers also had significant capacities in project management and often in engineering. Scale and the ability to plan a pipeline of projects over a relatively long period meant that the large power producers were able to sign multi-year framework agreements with turbine suppliers, with contracts foreseeing often over 1GW in capacity.

The increasing size of these contracts allowed big turbine producers such as Vestas, Gamesa and Siemens a higher degree of visibility in terms of revenues, and allowed them to ramp up investments and production. Examples include the 1.15GW deal that Siemens signed with German utility E.ON in 2008; the 2.1GW framework agreement that Vestas signed with Portugal's EDPR in 2010; agreements between Enel Green Power and both Siemens and Vestas for a total of 2.6GW; and two overlapping framework deals signed between Gamesa and Iberdrola in 2008 and 2011. Denmark's DONG Energy signed the biggest-ever offshore framework in 2009, for a total of 2.07GW of Siemens turbines.

In turn, the increasing scale of power producers contributed to bringing down the cost of energy (CoE) of wind in a number of ways. First, their investment

Table 2.2 NextEra's wind-power installations

Year	Capacity (GW)
2008	5,388
2009	6,493
2010	7,624
2011	8,386
2012	8,887
2013	10,117

helped push up the average size of wind farms, which studies suggest reduces cost – at least to a certain point. Second, the big producers were able to use their purchasing power to push down or at least limit turbine prices. And third, by establishing control centres that processed raw performance data from turbines – something which had previously been the jealously guarded preserve of OEMs – the large producers were able to introduce a higher degree of transparency into the market.

At Iberdrola's CORE (Renewable Energies Operation Centre) in Toledo, staff sit watching giant monitors, where they can monitor and control wind farms around the world, including being able to look 'inside' individual wind turbines. CORE director Gustavo Moreno Gutiérrez told me during a visit that Iberdrola had worked hard to break down a culture of secrecy among wind-turbine manufacturers – Iberdrola uses at least nine different ones – and ensure that the company has access to all the information it needs to be able to ensure cost efficiency. 'We can compare all kinds of performance data and that allows us to push efficiency standards higher', he said, adding that until 'very recently', only the manufacturers 'had the key to turbines and the operators couldn't even get in'.

Negotiating with OEMs to make sure that access to the right information was provided was one part of the challenge, but dealing with the data was another. 'One of the difficult things was to get the right type of information and then integrate all the different types of data from different manufacturers using a common standard', says Moreno. Being able to compare availability across

Figure 2.5 Iberdrola's CORE wind-farm control centre (Source: Iberdrola).

turbine types means that performance gaps between different models and turbines of the same type narrowed noticeably, say officials from Iberdrola. Other companies, such as EDPR, report similar results.

The leading utility developers achieved spectacular growth for their portfolios during 2005–11, with the two Spanish developers Iberdrola and Acciona, NextEra in the US, and Portuguese-owned but Madrid-based EDPR in the fore-front and others such as E.ON investing heavily. Former Siemens Wind CTO Henrik Stiesdal points out that during the period when the financial crisis limited the flow of investment to smaller developers, the utilities were able to keep market growth going by financing wind farms from their own balance sheets.

In offshore, utilities played an even more fundamental role in building the sector, with companies like DONG Energy, E.ON, RWE, SSE and Vattenfall taking the lead. 'I am convinced that virtually none of the offshore capacity we have seen built over the last five years would have been built without the involvement of utilities', says analyst and consultant Robert Clover (personal communication).

According to Clover, utility/IPPs (independent power producers) account for around 35 per cent of wind-asset ownership in Europe – where smaller developers and community wind farms are prevalent – but a much larger share of new investments. In China, utilities and IPPs account for more than 90 per cent of asset ownership, while in the US they account for 75 per cent.

After the big growth spurt of 2005–11, growth began to slow for the Western utilities, as a number of factors like a lack of availability of power purchasing agreements (PPAs) in the US and the Spanish crisis hit home. Attempts to make up for the slowdown by expanding into new fast-growing frontier regions in Eastern Europe hit regulatory and infrastructure barriers. An example is Romania, where Iberdrola announced its huge 1.5GW Project Dobrogea – hailed as the largest onshore wind project in the world – in 2010. Iberdrola CEO Ignacio Sanchez Galán flew to Bucharest in September 2010 to meet with Romanian Economy Minister Ion Ariton in an attempt to ensure that the project obtained grid access, after receiving a permit from transmission operator Transelectrica a few months earlier (Backwell 2010a).

Three years later, after a series of disputes over land with Czech utility CEZ – a rival wind developer – and faced with seemingly insurmountable bureaucratic barriers to establishing the transmission infrastructure it needed, the window of opportunity had passed, as Romania realised it had built too fast at too high prices. In October 2012, after completing just one 80MW project in the country, Iberdrola CEO Ignacio Sanchez Galan effectively called time on Project Dobrogea.

The publicly floated vehicles that the utilities had established had suffered from a lack of investor appetite, particularly after Copenhagen, and their share prices consistently failed to reflect the value of their assets or development portfolios. This meant that, on the whole, they did not serve their purpose as vehicles for bringing additional investment into the sector, and Iberdrola and

EDF both moved to reincorporate their subsidiaries in 2011. In contrast, however, Italian utility giant Enel's Enel Green Power (EGP) subsidiary has thrived since being spun off in 2010, with a self-financing model based on steadily growing cash flow.

Stagnant or falling power demand in Europe, the effect on utilities' 'legacy' fossil-fuel generation assets of increasing quantities of renewable power in the grid, and big capital investment requirements driven by shut-downs of existing fossil and nuclear capacity put pressure on company balance sheets and brought debt downgrades from credit rating agencies.

Utility executives became increasingly strident in their appeals for support for their fossil-fuel operations, arguing that gas capacity in particular had to be supported to provide back-up for 'intermittent' wind and solar supply, as well as criticising solar PV generation in particular as high-cost. Even when continuing to invest in renewables, and with their renewables businesses often the most profitable among their divisions, utility bosses began to call for governments to slow down the development of wind and solar or even tax renewables.

In October 2013, the CEOs of ten European utilities, including big wind investors like Iberdrola, E.ON and Enel, held a joint press conference in Brussels where they claimed that renewables were wrecking the European power industry. 'European energy policy has run into the wall', GDF Suez CEO Gerard Mestrallet said. 'In sectors like steel, cars and refining, when there was overcapacity, capacity was closed. But in the energy sector, we have massively subsidised additional capacity in solar and wind, which has led us to the absurd situation in which we find ourselves today', Mestrallet said (De Clercq 2013).

As well as a reining-in of subsidies for renewables, the CEOs appealed for an EU-wide system that rewards back-up capacity, a reform of Europe's struggling carbon-trading system and the setting of 2030 emissions targets to underpin de-carbonisation. The CEOs seemed to have had at least partial success in lobbying for this agenda in the European Commission white paper that was published in January 2013. The white paper calls for an EU emissions target of 40 per cent by 2030, and a renewables target of a modest 27 per cent, which is furthermore not binding on individual EU countries. Renewables companies expressed disappointment in the plan.

It is still too early to say how the conflict of interest within European utilities will work out, and how they will end up lining up in debates on energy regulation in the coming years. Many of the utilities represented in Brussels have thriving renewables businesses, whereas others are more reliant on their traditional fossil and nuclear power businesses. There are also legitimate concerns amid the complaints about renewables. Dysfunctional power markets in Europe have had the perverse effect of pushing gas power generation off grid when the wind is blowing, while allowing an increased use of much dirtier coal (see Nicola and Bauerova 2014).

Certainly not all of the current protests from utilities about Europe's energy regulations come from the point of view of an unenlightened defence of the continent's legacy generation system. Michael Lewis, Chief Operating Officer for

Wind at E.ON, the German utility that has been one of the most active investors in the sector, told a forum I organised in Frankfurt:

> It's not that utilities – and certainly not E.ON – are against renewables. What we are saying is yes we support renewables, but we support them in the context of a market that functions properly and keeps the lights on. And that bit of the message is not getting through.
>
> (Remarks at Recharge Thought Leaders event,
> EWEA Offshore Conference 2013)

Ultimately, utilities will take their lead from politicians and regulators. Clear rules to progressively decarbonise electricity generation and unequivocal political signals that there will be no turning back will allow those utilities that moved first and decisively into renewables to successfully make a transition to a new business model. Others, however, particularly those most tied to dirty, coal-based generation, show every sign of wanting to sacrifice the struggle to restrict climate change to safe levels for their own narrow interests. Given their current balance-sheet problems, utilities may not be as dominant in Europe as they have been, but they will continue to be the biggest players, particularly in the offshore arena.

Like wind-turbine manufacturers, financing constraints have created considerable scope for consolidation in the utility–developer space, and led companies to seek investment from capital-rich Asia. One of the most notable signs of the changing landscape was sale of a 21.35 per cent stake in Portuguese utility EDP to China's Three Gorges China Corporation for €2.7bn (US$3.7bn). Three Gorges manages a huge portfolio of hydropower assets (including the mega-project of the same name), but it is also a major investor in wind farms, and an investor in leading Chinese wind-turbine manufacturer, Goldwind.

Under the terms of the deal, Three Gorges agreed to invest €2bn (US$2.8bn) in a portfolio of 900MW of operational and 600MW of ready-to-build projects by EDP Renovaveis between 2012 and 2015. Of these, 600–800MW will be in Europe, a similar amount in the US and up to 200MW in South America. The first €800m (US$1,000m) will be invested during the first 12 months after closing the agreement. 'We are going to combine efforts to become worldwide leaders in renewable energy through joint development and ownership of selected renewables projects', EDP CEO António Mexia says (Mexia 2011).

CTG Vice-President Lin Chuxue told reporters in Portugal: 'EDPR is an excellent platform for the future for Three Gorges', adding that it will help the group 'to arrive at other businesses and other markets'.

The deal made Three Gorges EDP's biggest shareholder, and the Chinese company also arranged a €2bn (US$2.8bn) credit facility provided by a Chinese financial institution for up to 20 years. 'This liquidity is very important, as everybody can imagine, and we believe that this deal has a very positive impact on EDP's credit profile', said Mexia (Mexia 2011).

In August 2013, Japanese trading and finance house Marubeni agreed to take a 25 per cent stake in pioneering renewables developer Mainstream Renewable

Power for a reported €100m (US$138m), in a deal that will also see Marubeni take equity stakes in Mainstream's projects. Officials from Marubeni said that they were particularly interested in Mainstream's widespread participation in UK offshore wind projects.

The deal came on the same day that Marubeni agreed to buy a 50 per cent share in the 3.3GW of Portuguese generation assets owned by GDF Suez – including wind and solar PV. The company had already purchased a 49.19 per cent stake in DONG Energy's 172MW Gunfleet Sands offshore wind project, as well as stakes in onshore wind projects such as EDF's 205MW Lakefield wind farm in Minnesota.

As European utilities and developers find their ability to finance big project portfolios increasingly under pressure, we can expect to see more deals with capital-rich Asian companies. Meanwhile, increased competition among turbine manufacturers wasn't just coming from established Western engineering giants. The wind industry was about to experience the China effect.

3   China shakes the wind industry

Across the arid plains of Gansu, an endless array of wind turbines stretches towards the horizon. The turbines are part of a series of 'large wind-power bases' that are growing across northern China, from Xinjiang in the West to Jilin in the East. The onshore wind bases had a combined capacity of 39GW (yes, that is gigawatts) at the end of 2012, and were programmed to reach 165GW by 2020. The scale of China's ambition is, as ever, breath-taking.

China's wind-energy industry started late compared to the US and Europe. But it didn't take long to ramp up. Annual installations in 2002 were 66.3MW, but they had risen to 1.29GW by 2006. Installations mushroomed to 13.8GW in 2009, 18.93GW in 2010 and 17.6GW in 2011. China overtook the US for the first time as the world's largest market by cumulative installed capacity in 2010, and by the end of 2016, had a cumulative capacity of 145GW compared to 82.7GW in the US.

Foreign turbine manufacturers, led by Vestas and Gamesa, had taken a lead in starting China's wind-turbine industry in the hope of gaining huge new markets for their products. Vestas in particular made important contributions to China's industry through activities like carrying out studies – at its own expense – of China's grid requirements. In the middle of the last decade, Western turbine suppliers including Vestas, Gamesa, GE, Suzlon and Nordex had carved out more than 70 per cent of the market.

In a scenario familiar to other industries, however, China's wind boom was accompanied by a rapid expansion in domestic wind suppliers. By 2010, the top Chinese companies accounted for around 90 per cent of the local market and were challenging established suppliers in the ranking of the top ten global turbine suppliers. In 2011, the then biggest Chinese supplier, Sinovel, which we will look at in detail in a moment, was challenging Vestas for the number one spot.

How did they do it? On the one hand, Chinese companies learnt fast. Often taking licensed Western designs for a base, they were able to build up large-scale manufacturing and execute large-scale projects at breakneck speed. And they were able to supply turbines at a price level that put severe pressure on the margins of the established Western players. Analysts estimate that prices have fallen by more than 40 per cent since 2008, as companies have expanded and the industry has attracted new entrants prepared to offer low prices in return for initial sales.

They were also helped by a series of rules that were aimed at favouring local content between 2005 and 2010 during and, as we shall see, were able to leverage central and regional government contacts on an ongoing basis. While Western companies were able to continue to win smaller contracts from utilities, the big wind concession projects were effectively reserved for Chinese companies only, allowing a number of local companies to achieve massive sales. Finally, Chinese turbine manufacturers and utilities were able to take advantage of almost unlimited funding from state-owned financial institutions, such as the China Development Bank, to set up factories and build projects.

The massive expansion of turbine supply implied a big compromise in terms of quality, the effects of which have yet to be fully played out. China saw a large number of industrial accidents involving turbines during 2009–12, including tower collapses, blades shearing off, fires and electrical accidents that saw a number of workers killed.

As the Chinese Wind Energy Association describes:

> In recent years, the disadvantages related to high-speed development began to reveal themselves. Starting from the second half of 2009, a series of incidents happened in succession, including tower collapses, blade ruptures, nacelle fires, personal electric shocks and engineering accidents, leading to more than ten casualties and damage to more than ten sets of generator equipment.
>
> (Li et al. 2012: 63)

Mechanical failure was so rife at some of the big northern wind farms that operators sometimes had as many as one person per turbine working on operations and maintenance (compared to a couple of technicians for a large wind farm in the West). Stories abounded of technicians sleeping in the nacelles of machines to stave off cold weather conditions.

What allowed such widespread quality problems to exist was the fact that large utilities were more concerned about reaching almost recklessly ambitious internal and government targets as quickly as possible, and at as low a cost per MW installed as possible, rather than considering the cost of energy per MWh over the projects' full life cycles. Indeed, it is debatable whether many of the projects that were installed in the boom years will really have the 20-year life cycles that are typically the base for calculating wind farms' returns and values.

China's existing power system was also finding it hard to digest wind power, partly due to the huge new volumes and partly due to the low level 'grid friendliness' of Chinese turbines. Since 2011, an increasing number of incidents of turbines tripping off the grid have occurred, with 193 incidents of this kind across the whole country during January to August 2011 alone, according to the CWEA. Analysis carried out by the State Electricity Regulatory Commission identified four major issues: a) most wind turbines have no low-voltage ride-through capability; b) there are many quality problems in wind farm construction; c) connection of large-scale wind farms could threaten the stability and safety of the power grid; and d) wind farm operation management is weak.

By 2011, China's wind industry was facing massive growing pains. The country's grid system was buckling under the strain of bringing on the new wind capacity, particularly because the bulk of projects had been located in the arid north and west, far from the densely populated – and transmission system-dense – demand centres in the eastern and southern coastal areas. According to the China Electricity Council, 47.84GW of capacity was connected to the grid, out of a total installed capacity of 66.4GW, while 10 billion kWh of wind power was not produced due to curtailment. Regionally, the amount of wind electricity curtailed in the north and northeast were the largest, accounting for 57.20 per cent and 38.33 per cent of the total wind electricity curtailed nationwide, respectively.

In the latter part of 2011, the market began to slow down fast. It soon became clear that a massive build in inventories was taking place, with OEMs struggling to place their production. It also became clear that China's wind boom and easy financing had attracted too many turbine manufacturers into the market. According to analysts, there were over 80 Chinese OEMs in 2010 and around 70 in 2011, although many of these never actually entered the market in earnest.

According to BTM Consult, manufacturers collectively had around 28GW of annual capacity in 2011, compared to annual installations of 17.6GW in 2011 and 12.96GW in 2012. Although the level of installations was still the second highest in the world – as we have seen the US bounced back into first place with a record 13.2GW – 2012 became 'the year of adjustment' for China's wind industry. Many smaller turbine manufacturers simply stopped producing and shifted their activities back into other industrial areas, while, as we shall see, the adjustment had differing effects on the fortunes of companies within the top 15 companies, which accounted for 93.9 per cent of China's total installations in 2011.

The government rushed to build a series of major long-distance transmission lines, connecting the northern areas to the demand centres, while encouraging growth closer to demand centres, and slowing down the development of the big northern wind bases. As the Chinese Wind Energy Association says:

> On the one hand, the government was looking into different options to increase the consumption of the wind in [the producing] regions, as well as to increase the transmissions to the neighbouring regions with higher electricity demand. On the other hand, the government also realised that before the problem of electricity consumption or transmission can be solved overnight, it would be good to slow down the process of Wind Bases while starting to develop wind in the central and east area, where wind resources may not be prominent but electricity load is higher and transmission infrastructure is robust.
>
> (Li, et al. 2012: 13)

This sent the surviving wind-turbine manufacturers scrambling to adapt to a new type of market. First, developers were now in more of a position to pick and choose when selecting turbines, but new government regulations and increasing sophistication among operators meant in many cases a renewed emphasis on quality and cost of energy over project life-times.

Second, developers shifted their investment towards projects to transmission-rich areas closer to demand centres, which meant moving to sites that on average were less windy. The turbine manufacturers that were in a position to do so began to produce models with larger rotor sizes designed to capture more wind at lower speeds.

Third, manufacturers began to step up efforts to place production in international markets, as part of an ongoing 'go global' strategy encouraged by the government. Chinese companies like Sinovel had already become a fixture at international trade shows with impressive pavilions, and companies attempted to close numerous deals, from Africa to Eastern Europe to the Americas. There was continued – although unfounded – speculation through 2011 and much of 2012 that a Chinese company could be preparing a major acquisition of a European wind-turbine producer, even on the scale of Vestas, with Mingyang among the names mentioned.

It is interesting to look at the rollercoaster ride of the Chinese wind-turbine industry through the stories of two companies; Sinovel, which was the fastest-growing wind-turbine company in the world during the boom years and briefly challenged Vestas for global number one spot; and Goldwind, which is the current number one in China.

Sinovel's rise and stall

The meteoric rise of Sinovel was one of the most visible products of the boom in the Chinese wind industry. In 2011, just five years after the company was set up, it had become the world's second-largest turbine manufacturer – with 4.4GW of annual installations and 11 per cent of the global market – and was threatening to overtake the leader, Vestas. Its initial public offering (IPO) that year was the most expensive ever on the Shanghai stock exchange, with shares being listed at 90 yuan (US$14) each.

By 2013, however Sinovel's market share had plummeted, its shares had fallen to less than five yuan; it was being sued by former components supplier AMSC for US$1.2bn and being investigated by Chinese authorities for suspected violations of securities laws, in a case that reportedly involved inflating its revenue numbers (see Lee and Publicover 2014).

The company reported a net loss of 3.1bn Yuan (US$494.4m) in 2016, marking its fifth straight year of red ink. The loss was due to a decline in total installations, delayed orders and postponed payments, it said in a statement to the Shanghai Stock Exchange (SSE). The slump in Sinovel's fortunes marked a precipitous reversal of fortunes for a company that had been close to reaching the pinnacle of the global wind-turbine industry.

Sinovel was founded in 2006, and started life as an electrical-equipment subsidiary of state-owned industrial group Dalian Heavy. With no prior experience of manufacturing wind turbines, Sinovel licensed technology to produce a 1.5MW turbine from German turbine manufacturer, Fuhrländer. Less than two years later it had been awarded a huge 1.8GW order for the first phase of the Jiuquan wind park. 'To go from producing 50 units one year, to 1,000 units two years later was impressive', an industry source told *Recharge*. 'A European firm would never have expanded this fast.'

The key to Sinovel's growth was a perfectly timed product offering – no other Chinese company had a 1MW+ size turbine on the market – while parent company Dalian Heavy had plenty of factory capacity and engineering expertise.

Sinovel's then Chairman and President, Han Junliang, was extremely effective at marketing the company to investors, customers and policymakers. Junliang had an impressive array of high-level political connections, and his unprecedented efforts to nurture these relationships left others floundering in Sinovel's wake. For instance, a subsidiary of New Horizon Capital – co-founded by Wen Yunsong, the son of Chinese Premier Wen Jiabao – bought a 12 per cent stake in the company. To shore up his government connections, Han reportedly lavished gifts on officials, typically flying expensive fresh seafood from Dalian to Beijing for the Chinese New Year, or handing out US\$1,000 bottles of whisky on other occasions.

Sinovel enjoyed a strong connection with Zhang Guobao, the former Vice-Chairman of the National Development and Reform Commission (NDRC), the country's powerful policymaking body. In 2004, Zhang Guobao was tasked with overseeing the restructuring of heavy-industry groups in northern China. 'He had a mission to support all heavy industry to do something new, to re-establish growth', said a local industry participant. 'This [Sinovel's founding] was certainly a project in line with government policy' (Patton, Dominique, The Rise and Stall of Sinovel, Recharge 6 January 2013).

Zhang was a strong supporter of 'national champions' and everything suggests that Han knew he would win government support for Sinovel's larger turbine. The state-owned power industry is 'sensitive to the country's development', a former Sinovel employee explained to trade magazine *Recharge*. 'If a local producer can make bigger turbines, they will buy them. They want to support local production and it makes the government look good to have big turbines.'

In 2008, Zhang became director of the newly created National Energy Administration, just as China launched its first round of concessions for the mega-wind bases such as Jiuquan. The government did not specify a requirement for Chinese turbines, but no foreign firms won significant orders from these tenders.

Of the local players, Sinovel took the lion's share. After coming away with 1.8GW of orders in the 2008 tender – close to half the total – it followed up with 2GW from the Inner Mongolia and Hebei concessions in 2009, and a further 1.35GW the following year. These concessions played a large part in creating Sinovel's huge order backlog, and fuelled Han's ambitions for further growth. By

UNITED NATIONS CLIMATE CHANGE CONF

Figure 3.1 Sinovel's Chairman and President, Han Junliang (Source: Christoph
        Bangert/Danish Wind Industry Association).

the end of 2010, Sinovel had orders in hand of more than 14GW. Han told
investors ahead of the company's IPO that Sinovel would see a compound annual
growth rate of more than 30 per cent for the next five years. Indeed, its sales had
surged by 48 per cent during 2010.

But, around the same time as Zhang's retirement in early 2011, the 'year of
adjustment' hit, and Sinovel's management realised it had become too
dependent on large government tenders and a handful of key customers.

Many of China's large concessions were taking longer than expected to be
built, and Sinovel's inventories were piling up. Realising it had vastly overestimated its expected output, in March 2011 Sinovel abruptly refused component
shipments worth US$70m from its key converter supplier AMSC. This dealt a
body blow to AMSC, which subsequently sued Sinovel in Chinese courts for
these unpaid deliveries, as well as a further US$700m for unfulfilled contracts.
Worse, Sinovel's involvement with AMSC, which was co-designing turbines
with the company as well as supplying components, also led to an industrial
espionage case that would rock the international wind industry, and cripple the
Chinese company's plans to become a global player.

According to the allegations from AMSC, which lodged three civil suits in China, seeking a total of US$456m in compensation for intellectual-property theft, Sinovel stole software codes needed to upgrade its control systems to meet new grid standards in China. In September 2011, an Austrian court sentenced former AMSC Windtec employee Dejan Karabrasevic to a year in jail for 'fraudulent misuse of data' after concluding that he had passed the company's software codes to Sinovel in exchange for €15,000 (US$19,500). The evidence suggests that Sinovel was preparing to replace key components supplied by AMSC with locally manufactured equivalents, and needed to crack the software codes in order to do so.

Global media coverage of the lawsuits has hit Sinovel's international reputation, scuppering its largest overseas deal to date – a 1GW contract with Ireland's Mainstream Renewable Power – and destroying any chance of making sales in the US. 'We are all bound by law in Europe, and I suspect that I could be put in prison for receiving stolen goods', Mainstream founder and CEO Eddie O'Connor told reporters in Amsterdam in November 2011, shortly after cancelling the deal. He added that he had been doing everything he could to persuade Sinovel to take part in arbitration, and that Mainstream would be keen to work with the Chinese company in the future.

In Brazil, the threat of legal action from AMSC led to developer Desenvix filing a court order against Sinovel, demanding the right to inspect the 23 turbines it had ordered, to ensure they did not contain intellectual property 'stolen' from AMSC. (Desenvix dropped the case after it emerged that the turbines used control systems from US firm Emerson Electric.)

The Mainstream contract loss came as the firm's domestic orders were also shrinking. Large concessions had ground to a halt following a series of major accidents at wind farms in Gansu and Hebei that raised concerns about the safety of Chinese turbines. 'Sinovel benefited a lot from the large wind bases, but they tended to get larger volume orders from fewer customers', said Justin Wu, a wind analyst at Bloomberg New Energy Finance. 'This means they have been more affected by the slowdown in the north. And they had exposure to fewer clients than Goldwind, for example' (cited in Patton 2013).

The toll from a torrid 2011 was revealed in Sinovel's full-year report. Sales had dropped by almost 50 per cent, while profits were down 73 per cent to 776m yuan (US$125m). The numbers significantly worsened in 2012, with a loss of 582.7m yuan (US$94.5m). The company was also losing market share rapidly, falling to third in the China rankings – behind Goldwind and United Power – in 2012 and much further down the rankings in 2013 with a China market share of less than 6 per cent.

Sinovel's Vice-President Tao Gang described the market environment for the company from 2012 onwards as 'very tough'. Referring to falling turbine prices and the prospect of delayed payments from customers, he said: 'On one side is the volume, and on the other side is your profitability. The more you install, potentially the more money you lose, either currently or in the future' (cited in Patton 2013).

One symptom of the change in Sinovel's position was that orders from developer Huaneng, its largest customer, almost entirely ceased. In 2010, Huaneng bought about 40 per cent of Sinovel's turbines, according to data compiled by Beijing-based consultancy Azure International. In 2011, it consumed only 28 per cent of Sinovel's reduced delivery volume, and orders fell even further in 2012 and 2013. 'In the beginning, there wasn't much choice of suppliers for 1.5MW turbines so we bought a lot of Sinovel turbines', Jiang Xueyong, investor relations manager at Huaneng Renewables, told *Recharge Magazine* (*Recharge* 2013a).

But Huaneng placed few orders with Sinovel in 2012. 'Last year, we were building most of our projects in the south on high plateaus, so we were looking for turbines suited to those conditions. We've tended to buy a lot of turbines from CSR [China South Railway's wind unit].' In a widely reported speech at the 2011 China Offshore Wind conference in Shanghai, Xie Changjun, the president of China's leading developer Longyuan, attacked Sinovel for its poor quality record.

The question of quality and safety has also negatively affected Sinovel's sales. In a report to investors in September 2011, Daiwa Securities analysts in Hong Kong concluded that about 40 per cent of the 27 wind-turbine accidents they counted during 2010 had involved either Dongfang or Sinovel models. More accidents occurred in 2011, including electrical accidents that killed workers, as well as a disastrous collapse of a crane at Sinovel's Jiuquan factory that killed a high-ranking local communist official.

In March 2013, Sinovel brought in a former director of the Shanghai stock exchange and long-time Sinovel shareholder, Wei Wenyuan, to replace Han Junliang. Wei started by closing three facilities – the Dalian factory, a sales and service centre in Jiangsu Nantong and a factory under construction in southern Guizhou province. He also put 400 staff on 'paid holiday' – mostly from the research and development (R&D) department. Two months later, though, Wei was out.

Han – a one-time billionaire thanks to his 12 per cent stake in the company – was voted out of his role as company president at an August 2012 board meeting and subsequently stepped down "for personal reasons" as Chairman in March 2013.

To add to Sinovel's terrible 2013, Chinese authorities initiated an investigation into the company in May, reportedly for inflating its revenues in its stock-market filings. The probe was still under way at the time of writing, and had reportedly turned into a criminal investigation.

Sinovel appears to have significantly cut back its ambitions, such as building a 10MW turbine and becoming the world's leading turbine manufacturer within five years of its stock-exchange listing. It also seems to be set to fall further down the rankings of Chinese turbine manufacturers by market share, as companies like United Power, Ming Yang, SeWind and XEMC continue to challenge, but it still has significant strengths.

Sinovel has a significant position in China's nascent – but fast-growing – offshore market. At a conference in Beijing in October 2013, Vice-President

Chen Danghui said 'We've received lots of negative coverage, but we can overcome (these difficulties) because our core people are still around.' The company is working hard to fill the gap left by AMSC, working with Danish control electronics provider Mita-Teknik, and third-party testing agencies GL and TÜV Nord to certify its turbines, a key step in ensuring credibility in international markets.

With further big financial losses in following years, stock market delisting, and the legal dispute with AMSC unresolved as the end of 2016 approached, Sinovel was facing a hard slog to get out of its problems, and would find it hard to re-establish its position in the face of now much stronger competitors. Finally, in late 2016, a Chinese court in Hainan dismissed one of the civil suits, with AMSC announcing that it was intending to appeal to China's Supreme People's Court (see www.sec.gov/Archives/edgar/data/880807/000119312516688767/d226162d8k.htm).

According to data from FTI Consulting, the former number one manufacturer in China (and number two globally) was in 18th place with just 1.0 per cent of market share in its local market in 2016.

Goldwind plays the long game

In contrast to Sinovel, Xinjiang-based Goldwind Science and Technology has managed to maintain its market share and avoid major corporate upheaval and legal controversy. While its growth was not as spectacular as Sinovel's, Goldwind enjoys some notable advantages, in particular because it has significant design resources – and owns its own intellectual property rights – through its acquisition of German turbine manufacturer and designer, Vensys.

Wu Gang, an early pioneer of Chinese wind-power technology who worked on some of the first demonstration wind farms, was instrumental in setting up Goldwind, the country's first wind-turbine company, in 1997. By 2006, Goldwind controlled around 33 per cent of China's total market. It became a listed company in 2007 in a US$216m IPO on the Shenzhen exchange, and this put it in a position to acquire a 70 per cent stake in Vensys in 2008. It carried out a further IPO on the Hong Kong exchange in 2010m, raising US$917m.

Goldwind had begun a transition from gearbox-based technology to a permanent-magnet direct-drive (PMDD)-based design in 2003, when it partnered with Vensys to co-develop the GW1.2MW model, which subsequently evolved into the 1.5MW model. Since the acquisition, Goldwind has developed 2.5MW, 3MW and 6MW machines.

As well as designing Goldwind's machines, Vensys has licensed successful direct-drive designs to companies such as Impsa in Brazil and Regen Powertec in India. Goldwind has also been unique among Chinese manufacturers in carrying out a steady expansion into the 'mature' US market, as well as Australia and other countries.

Wu Gang – now Chairman – often emphasises the importance of a truly international technological community around wind, from his earliest visits to the Riso wind-research centre in Denmark. He told me in Germany in late 2012, 'We have an open mind and we try and learn from different countries' cultures, so that is quite different from other Chinese companies' (Wu Gang, personal communication).

Wu says that he sees a 'global' approach as key to his company's strategy, and says it encompasses five elements: internationalising the company's technology and products; internationalising its human resources; expanding into international markets; gaining access to international capital; and adopting an international management system.

A key area for Goldwind's international expansion is the US. In contrast to other Chinese turbine manufacturers that have announced big projects in the country which have then failed to materialise, Goldwind's emphasis has been to construct – and help finance – mid-sized projects such as the 109.5MW Shady Oaks project in Illinois in order to establish a track record, so that developers and utilities can see the effectiveness of its turbines. 'We have good communication with local customers, and we can help on the investment side of the project so they can easily see the performance', says Wu.

Goldwind is also aggressively seeking to expand in Australia, where it has set up a Sydney office; in Latin America, where it has secured projects in Chile and Panama; and in Pakistan. It is also leveraging the relationship between major shareholder Three Gorges and the Portuguese wind-development giant EDPR to jointly take part in tenders with the latter.

Goldwind's plan had been to dedicate around one-third of its production to international markets by the end of 2015, with major growth expected to come in other Asian countries, Africa, North and South America, and Australia. However, intense competition in global markets means that in common with other Chinese OEMs 'Go global' went slower than officials had expected (see below). Plans for a US factory were put on hold, and Wu says the company has not taken any firm decisions about setting up assembly operations outside China, although he states proudly that Goldwind US uses more local components than other 'local' manufacturers.

Like others Goldwind had to change its focus in order to get through the 'adjustment' years of 2012–2013. Goldwind was quite quick to see that the Chinese industry was about to go through major changes, and began to change its plans in 2011 to adapt to a period of intense competition and a reduction in the pace of expansion. 'We know it will be a very cold winter for the wind industry so, since last year, we have been preparing how to spend the wintertime', Wu told me in Husum, Germany, in late 2012. 'Sometimes winter can be good for companies to improve the quality of their management.'

In contrast to some of its Chinese competitors, Goldwind has put a continuous effort into R&D, ensuring that its technology keeps up with the latest international level. At the same time, it has strictly controlled employee numbers at around 4,000 employees, making it a very lean company for its size in terms of MW delivered. In 2012, it also outsourced its blade-manufacturing activities, selling its Tianhe subsidiary while maintaining a stake.

The company has also become more 'vertically integrated' by expanding its activities down into project development, financing and engineering procurement and construction (EPC) services – all key activities in a period of market slow-down – which make the company's turbines more attractive to potential buyers.

Similarly, Wu was confident of the tremendous strength that the Chinese wind market can provide to his company when faster growth resumed, pointing to the strong underlying support from government policymakers, and its access to large-scale loans from the China Development Bank to finance its expansion. As we shall see below Wu was right to be optimistic.

Then as now, Goldwind has been focused on organic growth and is not in the market for acquisitions of turbine suppliers, despite bouts of speculation. 'We are focused on new technology', he says. 'We don't want to buy companies with traditional technology. We are looking at the future.' Goldwind's roots in the remote northwest region of Xinjiang have taught it to keep an open mind, seek cooperation and expand cautiously, he says. 'Our approach is to grow step by step. We are different from some companies that see everything as simple', Wu says.

Adjustment leads to recovery

After the two years of adjustment in 2012 and 2013, the Chinese wind-turbine market resumed a faster pace of growth in 2014, surpassed the 20GW of installations a year mark, and then hit a new high in 2015. This did not make things any easier for turbine manufacturers, given the intense price competition between a leading group of around 10 companies, and continued overcapacity in the market.

As we have seen, Goldwind has been more successful at maintaining market share and corporate stability in tough times, and is the undisputed number one in China by market share. Others, such as Sinovel and Dongfang have seen themselves fall sharply down the rankings.

Although we have talked quite a bit about quality and intellectual property rights, the main factors in the fortunes of Chinese turbine manufacturers have been relationships with the 'big five' wind developers.

One notable trend has been the growth of state-owned turbine manufacturers with close links to utilities. One of the big winners in the turbine market in the early part of the decade, was Guodian United Power, which became the second-largest Chinese OEM in 2012 and achieved a market share of almost 10 per cent in 2013. Guodian of course is also the owner of Longyuan, China's largest developer. The company has enjoyed a strong presence in the market due to its relationship with Guodian, and developed a number of new turbine models including a 6MW offshore machine.

Shipbuilder CSIC is another company that has prospered and risen in the rankings, while other state-owned power companies such as Datang and State Grid have bought or set up their own turbine companies.

As well as the state-owned players, several other companies have taken advantage of the relative decline of Sinovel and Dongfang. These include NYSE listed Ming Yang, which has grown aggressively and is developing a series of innovative turbines with German designer, Aerodyn; the highly regarded Envision, XEMC and Shanghai Electric (which formed a joint venture with Siemens in late 2011).

Consolidation has been slow to occur in China, but as we will see, the market has become highly concentrated, and strong pack of leaders has emerged including a top three of Goldwind, Envision and United Power. As in other markets, China's wind sector is maturing and becoming more technically sophisticated, with an emphasis on the cost of energy and squeezing production out of existing turbines as well as constructing new capacity. The companies that are able to develop in terms of sophistication of products and services – the area of providing operations and maintenance services for China's growing fleet of turbines – and that are able to extend turbine life cycles will be in a position to gain market share and grow in the next coming period.

To some degree, the outcome of the battle will depend on turbine quality and good business strategy and management. But it will also depend to a great degree on key relationships with the big developers, as well as government industrial policy. While in other industry sectors like telecoms policy, committees have been proactive in forcing through consolidation, government policy as regards turbine manufacturing has been relatively hands off, with officials intervening at the 'macro' level of energy pricing and letting price competition largely decide who prospers or goes out of business.

The foreign OEMs try to hang on

Meanwhile, what is the situation for foreign manufacturers in China, and what are their chances of survival?

As we saw, Western turbine makers were quick to see the huge potential of the Chinese wind sector, and had carved out more than 70 per cent of the market by the middle of the last decade. By 2012, their market share had been eroded to only 9.8 per cent, according to BTM Consult. The top four non-Chinese manufacturers in China together had installed about 1GW in 2013, out of a total market of around 13GW, according to Justin Wu, lead wind-energy analyst at Bloomberg New Energy Finance in Hong Kong.

The reasons for the decline are myriad. Patrick Dai, a Hong Kong-based analyst at Macquarie Securities, puts the rise of Chinese OEMs down to 'a combination of competitive pricing and advanced technology, flexible customisation, easy servicing and relationships with the local governments' (Backwell and Publicover, The future of foreign firms in China, Recharge, 04 October 2013). Competition has been intense, particularly given overcapacity among the Chinese players, and the hiatus in the market during 2011–12 has led to intense pressure on margins.

In the past few years, a number of Western companies have decided that the Chinese market is not for them.

In October 2009, Spain's Acciona sold its 45 per cent stake in its joint venture with a subsidiary of state-owned China Aerospace Science and Technology Corporation (CASC) – just three years after establishing a plant in Nantong, which it claimed at the time was China's largest turbine plant.

In 2012, Indian-owned Suzlon – once a top-ten player in the market – announced that it was selling its manufacturing unit in China, despite top officials claiming just a year earlier that it was able to compete with local companies on costs and pricing. The sale of a 75 per cent stake in Suzlon Energy Tianjin to Poly LongMa Energy (Dalian) was eventually completed in September 2013, after an earlier deal with China Power New Energy – announced the previous June – fell through. Suzlon subsidiary, REpower, announced it was leaving the Chinese market in September 2011 because of 'increasing trade protectionism', and closed its majority-owned factory in Baotou, Inner Mongolia.

In July 2013, US manufacturer GE ended the joint venture (JV) it had formed with Chinese power-equipment group Harbin Electrical Machinery in September 2010, although it says it remains committed to the market.

Meanwhile, others have discreetly shelved plans to expand in China. These include Nordex – which tried unsuccessfully to conclude a joint venture with local developer Huadian in 2012, and subsequently scaled down its local operations – and Alstom Wind, which had planned to set up a new Chinese plant.

Those remaining say they are determined to stay, arguing that as the Chinese market matures – in terms of customer needs, geographical diversification and pricing, with widespread consolidation expected among local players – opportunities will begin to grow again. 'Once the transmission bottleneck is solved the market will take off again', says FTI's Feng Zhao. Given the market is so large, 'if the Western companies can just maintain their market share where it is, it's not so bad', he says. 'If you don't have the patience and you are not there, you are losing out on the business' (Backwell and Publicover 2013).

For Zhao, the key to success in China is offering the latest technology, particularly as the market is becoming more sophisticated and projects shift south to sites with lower wind speeds. 'I think we can see across emerging markets that if you arrive late and with old technology it's not going to work', he says (Backwell and Publicover 2013).

The other key factor is relationships with the major developers. 'If you look at the data, you can see that Vestas and Gamesa both have at least two relationships each with major customers', he points out (Backwell and Publicover 2013). The remaining foreign OEMs – Vestas, Gamesa, GE and Siemens – have widely different strategies, while there is a question mark over GE's future.

Vestas

Vestas was an early front-runner in China, and had installed more than 4GW in the country by the end of June 2013. Its ranking in China fell from sixth place in 2010 to tenth in 2012, when it accounted for just 3.2 per cent of the market. But it has strong relationships with developers such as Datang, China Resource

New Energy and CGN Windpower. And it has secured repeat business from many of the companies it has supplied.

Its attempt to sell locally produced V52-850kW and V60-850kW machines was a failure, and it closed its Hohhot factory, leaving the 'kilowatt' platform in summer 2012. But Vestas has maintained its Tianjin factory, as well as its Beijing and Shanghai offices. And it is winning a steady flow of orders for its 2MW V90 and V100 turbines, including recent orders from Chinese gas group Hanas New Energy and Hebei Construction and Investment Group.

Vestas' turbines feature yaw backup systems to protect blades from the typhoons that frequently lash China's southeast coastline, where it has been a particularly strong player. It has long won projects in Fujian province, for example, claiming roughly 600MW of the province's 1.6GW installed capacity by late 2012. And it should be well positioned to win orders when offshore takes off along the eastern seaboard, as Fujian, in particular, wants to install 500MW offshore by 2015.

The challenge companies like Vestas face is capturing niche segments such as offshore, low-wind and high-altitude projects with their technologically advanced turbines, says Wu. 'When you're able to source from China's supply chain itself, you're not necessarily offering those products', he says. 'The question for foreign manufacturers, really, is whether or not they can introduce their new products into China ... you're not competitive on cost when you move into those turbine segments' (Wu, quoted in Backwell and Publicover 2013).

In recent years, Vestas has also shown that it is prepared to scale down operations and stop participating in tenders if prices are not attractive, as it openly stated in 2012, as it subordinated its participation in China to the wider aim of returning the company to profitability. Since the difficult years of 2012–13, Vestas has continued to sell quantities of turbines that are modest within the context of the overall Chinese market, but nonetheless significant, using its technological advantage to good effect to supply projects in lower wind-speed areas.

It's market share has remained relatively stable, with even a slight recovery in 2016 to 2.2 per cent, and with the renewed buoyancy of its global sales and balances, there have even been reports of talks to acquire a major Chinese turbine maker.

Gamesa

Gamesa is widely tipped as one of the foreign OEMs most likely to continue to do well in China, mainly due to its strong relationships with some of the country's leading utility players, such as Longyuan and Huadian. It was ranked in ninth place in 2012 with 493MW, according to BTM Consult figures. '[We] can provide experience in niche markets and in specific technologies beyond what the local manufacturers can provide', said Gamesa China's boss José Antonio Miranda (Backwell and Publicover 2013), pointing to issues such as resistance to turbulence and extreme weather such as typhoons, and developing high-altitude sites. Gamesa is also looking at providing Operations & Maintenance (O&M) services, offering improvements in performance and the extension of turbine life cycles.

One key aspect of Gamesa's strategy is its approach to introducing technology into China, with an approach to intellectual-property (IP) protection that could be described as less obsessive than some of its peers. 'Two years ago, we decided to launch our products all at once in all regions', says Miranda. 'It's clear to us that to compete we have to come with our most up-to-date technology' (Backwell and Publicover 2013).

From the point of view of IP protection, Miranda describes China as 'no different'. 'In any market, your new product can be an inspiration, but we have not seen any IP theft here' (ibid.). Miranda notes that Gamesa has been producing in China since 1995, has a 'very developed' local supply chain and produces large turbine parts such as hubs that are sent to other parts of Asia and Europe. Gamesa is also eyeing the offshore market in the medium term, and says it would look to deploy its new 5MW turbine from 2016 onwards, although delays in the development of China's offshore wind sector has slowed the process.

In the future, Miranda estimates that there will be no more than 'two or three' foreign manufacturers operating in the market, but that there are 'no special difficulties' for non-Chinese players. 'The domestic companies will improve their quality and that has a cost, and the international companies will become more specialised in niche areas, and there won't be such big differences in domestic and international markets', he says (Backwell and Publicover 2013).

Like Vestas, Gamesa has managed to maintain a small but significant share of the Chinese market, amounting to some 2.1 per cent of the total in 2016.

Siemens

Siemens' strategy differs from Vestas' and Gamesa's in that the company has long said that it would focus on a JV with a local heavyweight industrial partner.

In late 2011, Siemens' plans came to fruition when it announced an agreement with long-term partner, Shanghai Electric. The two companies agreed to jointly invest €169.1m (US$226m) to set up two wind-power equipment joint ventures in China – Siemens Wind Turbines (Shanghai) and Shanghai Electric Wind Energy (Sewind) – with blade and nacelle plants in Shanghai producing Siemens' SWT-2.5–108 model.

The JV is producing turbines for onshore projects, including the 50MW Guangrao wind farm, 400km south of Beijing in Shandong province. An official from Shanghai Electric said in February 2013, when the project was completed, that the venture planned to install about 300MW of onshore wind in China that year.

Unsurprisingly, given Siemens' dominance in the sector, Sewind says that it hopes to claim around 25 per cent of China's offshore market by 2015.

Siemens supplied 21 of its 2.3MW turbines for Longyuan's Rudong intertidal wind project, which went operational in May 2012. Shanghai Electric had also won a tender in 2011 to supply 26 of its 3.6MW turbines to the 101MW second phase of the Donghai Bridge wind farm off Shanghai, and will supply the 200MW Luneng Jiangsu Dongtai project. Sewind says it expects to sell 75 of its SWT 4.0–130 turbines in 2014, and double that amount in 2015.

The logic of Siemens' move was clear, but there was scepticism about how successful the JV would be in practice. 'The question is, how many joint ventures have worked?' says Feng, pointing to the break-ups of GE and Harbin, Acciona and CASC, and others. 'Basically, it's a battle between technology and market', he adds. 'The Chinese want cutting-edge technology, but for foreign companies, the IP is the value and they want to hold on to it. It's a tough bargaining process' (Backwell and Publicover 2013).

Indeed, by 2015, following years of reported disagreements between officials from the two companies, Siemens and SEC announced that they were ending their joint venture in favour of a straight licensing deal, plus a supply chain agreement for the rotor blades to be produced at Siemens' factory in Shanghai, leaving the German company once again in a relatively marginal position in the Chinese market. It remains to be seen to what degree the recent merger of Siemens and Gamesa will significantly increase the company's weight and market share in the Chinese market.

GE

US turbine giant, GE, ended its joint venture with Harbin Electric in July 2013, less than three years after it was set up. But it is clear that the US company will never give up. Although several foreign suppliers have given up trying to make money in the Chinese market, GE decided to continue on its own. In 2014 it was ranked number 23 in China with only 0.23 per cent of the market. But it has improved its position slightly since then and was in 19th place in 2016 with 0.9 per cent of the market.

In 2012, GE ranked 17th in China, with 90MW installed, according to the Chinese Wind Energy Association, compared to 11th place with 408.5MW of deliveries a year earlier. According to Patrick Dai, GE does have a strong track record with east-coast projects, and the company was one of Longyuan's top five suppliers in 2012. However, Dai adds, 'I don't see any progress' (Backwell and Publicover 2013).

Wu believes GE's 'reasonably localised supply chain' (Backwell and Publicover 2013) could help it sustain its local operations. 'They're still quite active in China', he says (ibid.), noting that it is still winning new contracts and buying key components such as gearboxes from local suppliers, including China High Speed Transmission, one of the country's largest makers of wind-turbine gearboxes.

It also has the products and technologies to compete for low-wind and high-altitude projects in China. 'Certainly, GE's class-three wind-speed products have been quite successful in the West', says Wu (Backwell and Publicover 2013). As the acquisition of Alstom's power business has brought GE back to the offshore wind sector, GE has decided to re-enter the Chinese offshore wind industry (it had the plan to target the Chinese offshore wind market by setting up a factory in Jiangsu with Harbin Electric, but it failed) and the company has signed an agreement with Chinese Three Gorges Corporation to build its 6MW direct drive offshore wind turbine at the Chinese company's offshore wind industrial base in Fujian province in the end of 2016. Its recently acquired blade manufacturer LM Wind Power has also decided to produce in this offshore wind industrial base.

Boom: China's wind industry reaches new heights

By 2015, China's wind industry boom had once again picked up pace and installations were reaching levels that were dazzling the industry and attracting coverage from the world's main news outlets.

The main boost was a planned reduction in onshore wind Feed-in-Tariffs (FiT) announced by the National Development and Reform Commission in January 2015, which would see a reduction in tariffs paid to all projects in the top three wind regions commissioned after 1 January 2016, giving developers and equipment suppliers an incentive to rush their projects through. Wind-turbine manufacturers produced flat out to rush to meet delivery schedules.

First-half installation figures showed growth picking up sharply, but the full-year figures shook the world's energy scene. China installed some 31GW of wind in 2015, some 35 per cent higher than in 2014. Chinese installations accounted for almost half of total global installations. China now had a total installed capacity of 145GW, putting it firmly in the number one spot ahead of the US and

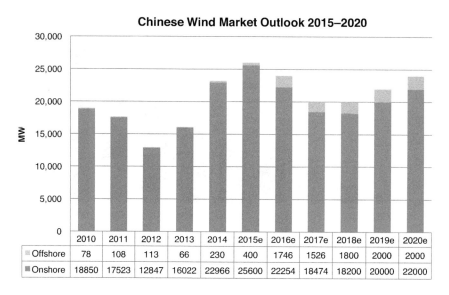

	2010	2011	2012	2013	2014	2015e	2016e	2017e	2018e	2019e	2020e
Offshore	78	108	113	66	230	400	1746	1526	1800	2000	2000
Onshore	18850	17523	12847	16022	22966	25600	22254	18474	18200	20000	22000

Figure 3.2 Chinese Wind Market Outlook 2015–2020 (Source: FTI Consulting).

the EU in its entirety, which had a cumulative capacity of 141.6GW of capacity. And for the first time, Chinese manufacturer Goldwind knocked Vestas off its top spot and became the world's largest wind-turbine manufacturer in terms of annual installations.

'For the wind industry, 2015 is the year of China', said Feng Zhao, a senior director in FTI's energy practice in an interview with the *Financial Times*. 'China not only helped make 2015 another record year for the wind industry but was home to five of the top 10 turbine manufacturers in the world.' (Clark 2016).

This does not mean that the development of China's wind market had put all of its problems behind it. The explosive growth of wind-power installations in 2015 put the already stressed Chinese grid further under strain and brought the total amount of unconnected wind-power capacity up to 19,180 MW by the end of June 2015.

In the first half of the year, 17.4 billion kWh was curtailed in China, nearly 1.5 times as much as in 1H 2014. Regions that suffered the most were Jilin, Gansu, Xinjiang and Heilongjiang where the wind-power curtailment rates were 43.0 per cent, 31.0 per cent, 28.8 per cent and 22.7 per cent respectively. Other areas with double digit curtailment rates were the Inner Mongolia autonomous region, Liaoning and Hebei province.

As we have seen, power transmission bottlenecks have stymied wind-power development in China since 2008 when wind-power installations started taking off. But the country is taking decisive measures to solve the problems. Its huge investments in building HVDC and HVAC lines are starting to pay off, and the amount of wind power connected to the power grid in 2014 reached 19,810 MW,

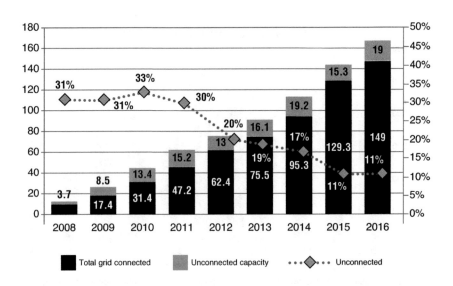

Figure 3.3 Unconnected wind-power capacity in China 2008–2016 (Source: FTI Consulting).

40 per cent more than 2013 and the highest rate China has ever had, although this was still not enough to prevent the absolute amount of wind power that was not connected to the grid rising to its highest level yet.

Further big projects that can transfer electricity generated in north, northwest and northeast of China to the load centres along the east coast and in central China are being constructed, however. China State grid was expected to invest another RMB420.4bn (US$67.6bn) in grid connections in 2015 alone. And meanwhile, the NEA's 'Notice for the Construction of 12 Power Transmission Lines for Speeding up the Work Plan for Prevention of Air Pollution' published in mid-2014 foresaw five of the 12 new transmission lines due to be completed in 2016 and the rest completed in 2017.

In the wake of the boom – who are the winners?

The boom accelerated important changes in China's wind market, increasing the dominance of the big turbine producers with leading manufacturers seeing growth rates of greater than 80 per cent compared to 2014. In particular, number one producer Goldwind has continued to expand its market share, while Envision has successfully challenged both United Power and Mingyang for second place. CSCI Haizhuang also saw strong growth.

But 2015 was an exceptional year because of the installation frenzy caused by the planned change in FiTs. Annual installations fell from 31GW to 23GW in 2016, and were expected to drop further to below 20GW in 2017 before staying flat in 2018/19, starting to rise again in 2020 and then reaching new highs sometime in the early 2020s. This has led to even more intense competition on price, and will lead to further shake-ups among turbine manufacturers.

Feng Zhao notes that, if we take a mid-term view, consolidation through M&A has been slower than expected in China. In 2016 there were 22 Chinese OEMs installing wind turbines in China, compared to 23 producers in 2014–2015, with one company going out of business. There was only one M&A deal in 2016, the acquisition of turbine manufacturer China Creative Wind Power (CCWE) by an industry group DunAn that owns a Tier 3 wind turbine manufacturer, Jiuhe Energy. But this masks a wider story of deepening concentration. Zhao notes that in 2016, the top 5 producers in China – or so called Tier One – accounted for 60 per cent of the market, while the top 10 accounted for nearly 85 per cent of the market. The smallest 12 local producers, meanwhile, accounted for just about 10 per cent of market share.

Zhao considers that the 2015 boom allowed some small producers with local niches to stay in the market, but the expected slowdown from 2016 to 2019 will see many of these struggle in the new price environment. And meanwhile, the biggest players have no incentive to acquire the smaller companies, meaning that consolidation through a slow series of company failures is the most likely scenario for the Chinese market. 'For consolidation you need a good reason', he says. 'Some of the big players like Goldwind and Envision have their own R&D, intellectual property and global footprint. Those companies have no incentive to

Figure 3.4 Leading Chinese wind-turbine manufacturers (Source: FTI Consulting).

acquire those that don't have their IP or those who do not even have reliable turbine technologies.' he adds. The big companies also have already proven, reliable turbine offerings and good relationships with local utilities, component suppliers and financiers, making acquiring companies further down the food chain unattractive.

Meanwhile, three of the foreign companies have hung on in the local market and even slightly increased their market shares, with Vestas accounting for 2.2 per cent, Gamesa 2.1 per cent and GE 0.9 per cent – a total of 5.2 per cent in 2016 compared to 2.7 per cent in 2015. This is mainly due to the shift in turbine installations because of grid constraints from the windy northeast to the less windy areas of the south and centre of the country, which favours the technical sophistication and offerings of the Western Companies. However, this is a temporary situation and Western companies will find it hard to continue to increase market share as connectivity increases and prices continue to fall. Significantly, there have been persistent rumours that Vestas is in talks to acquire or merge with one of the top Chinese producers.

'Go Global' – postponed not cancelled

The idea that Chinese wind-turbine companies were about to move into international markets and drive incumbent manufacturers out of the market was almost an obsession among many observers of the wind industry during the first part of this decade. Financial analysts trying to take a mid-term view of wind stocks, in particular, were drawn by the idea, particularly as Chinese companies were effectively wiping out incumbent European and US companies in the solar PV sector in this period. However, by 2013 it was clear that the predicted wave of Chinese turbines flooding markets and widely predicted large-scale acquisitions by Chinese companies had not materialised. 'The go-global strategy has failed', says Eduard Sala de Vedruna, Senior Director at IHS Markit.

The explanations for why Chinese expansion has not happened as quickly as some expected fall into two categories. On the one hand, Go Global has not happened for turbine manufacturers because their largest clients – big Chinese developers like Longyuan and Datang – have been slower than expected in carrying out their Go Global plans, as they discover the difficulties of operating in a foreign environment, and often see relatively higher returns from local projects.

On the other hand, it was clear that most of the Chinese turbine companies were not adequately prepared for operating internationally. The most serious issues were those of the lack of a track record and the dubious quality reputation of some Chinese machines, coupled with the fact that many of these had not been certified by international companies like DNV or Garrad Hassan. In 2011, a time of intense Chinese activity in the US, developers complained that some Chinese turbine manufacturers did not have full documentation in English or wanted provisions for dispute resolution to be based on legal jurisdictions. 'Culture' was also a problem. Leading turbine companies like Sinovel often had enormous stands at international trade shows, but had staff who were unwilling or unable to engage with prospective customers when they approached.

There were some high-profile failures, such as a project to build a big US wind farm and manufacturing plant – with high-profile political support – by Chinese company A-Power. But it has been the Sinovel case that has shown what can really go wrong, with the AMSC dispute killing the biggest Chinese turbine deal outside China, and a number of other failed projects.

None of this means that Go Global is not happening, however. Companies like Goldwind have shown they can build successful projects in the US and elsewhere, and along with other Tier One Chinese players, they are competing to win more.

With the slowdown of installations in the Chinese market, the need to expand internationally has increased sharply in order to use their spare industrial capacity. Internationalisation is closely aligned with central government policy. The Chinese government is giving a major boost to the internationalisation of the sector through its 'Belt and Road' initiative.

'Belt and Road' refers to the Silk Road Economic Belt and the 21st Century Maritime Silk Road. The initiative was introduced by Chinese President Xi Jinping in 2015 against a background of (relatively) slower Chinese economic growth and the challenge of manufacturing overcapacity for domestic industries – something which continued to be a feature of the Chinese wind-turbine manufacturing sector. The plan has the goal of reviving the centuries-old trading routes linking Asia, Europe and Africa.

Elements of the plan that are likely to help boost the wind-power business include the connectivity of transport and energy infrastructure, the promotion of cooperation in environmental protection industries, as well as the promotion of in-depth cooperation with other countries along the 'Belt and Road' in new energy technologies. In parallel, the Chinese Government has set up a $40bn Silk Road Infrastructure Fund to provide financial support to assist Chinese firms in entering markets along the Silk Road. In addition, to address regional infrastructure bottlenecks and capital constraints, China has established the Asia Infrastructure Investment Bank (AIIB) along with 56 other countries, with authorised capital of US$100 billion.

The top three regions favoured by Chinese wind turbine manufacturers so far have been be Europe, Africa and South America. Since the 'Belt and Road' initiative will boost the transport of infrastructure construction and the

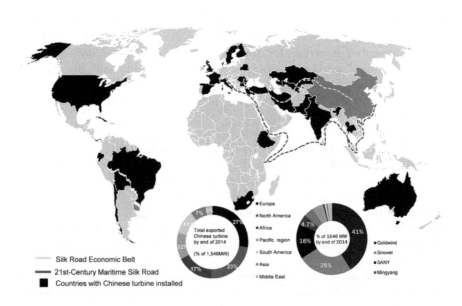

Figure 3.5 China's 'Belt and Road' initiative set to boost wind turbine exports (Source: FTI Consulting).

connectivity of energy infrastructure linking Asia, Europe and Africa, it is likely that Europe and Africa will remain important overseas markets for Chinese turbine manufacturers. But we can also expect more Chinese wind turbine exports to countries in Central Asia and South Asia (starting with Kazakhstan and Pakistan). Although South America is outside of these new trading initiatives, Chinese companies can be expected to be supported doing business there by receiving financial support from the Silk Road Infrastructure Fund and the AIIB.

Chinese OEMs – with proven and certified wind turbine technologies, overseas experience and strong ties with state-owned EPC contractors/developers involved in energy infrastructure development abroad – can be expected to gain more from the initiatives than smaller less established companies (Feng Zhao, China's "Belt and Road" initiative to boost Chinese wind turbine exports FTI Spark Note, 30 July 2015).

But this expansion of Chinese companies is unlikely to occur in the way that commentators and analysts expected back in 2011–12. The idea of Chinese turbine manufacturers taking over the world has perhaps been overblown – GWEC's Steve Sawyer calls it an expression of Sinophobia, and says 'The notion that the wind industry was going to follow a similar pattern to what we've seen in the solar industry, it was never going to happen – that was nonsense, made up by people who don't understand the industry.'

FTI Consulting's Feng Zhao says that it is more likely that Chinese companies expand internationally through acquiring projects than through out-and-out acquisition of turbine manufacturers. "The top Chinese companies like Goldwind and Envision have their own turbine IP so why would they want to spend a lot of money acquiring existing companies?" he asks. He adds that the extremely advantageous cost structure of the Chinese industry means that in most cases companies can export much of their equipment from China directly rather than set up local factories and still compete.

Unlike five years ago, Chinese turbines now have a long track record, having performed well in international projects in places as far flung as Chile, South Africa and Ethiopia meaning that turbines are 'bankable'. And they have strong support from government and government linked financial institutions.

However, Goldwind's Chairman and founder Wu Gang is still modest about Goldwind's ability to compete head to head against the big Western companies around the globe, saying in 2015, 'We are number one in the world in terms of market share, but we are well aware that we still lag behind multinationals like Siemens, GE and Vestas.' (Shepherd and Hornby 2016).

As Feng Zhao says, the large-scale acquisition of projects is likely to be the key to the expansion of Chinese companies in world markets. Chinese utilities have led the way in some countries. In 2013, Longyuan won the rights to build two wind projects in Northern Cape Province, South Africa, along with local company Mulilo Renewable Energy. The projects, De Aar phases 1 and 2, have a combined capacity of 244MW and use United Power turbines. Longyuan had previously won rights to develop the 100MW Dufferin wind farm in Ontario,

Canada, its first project outside China. China's State Power Investment Corporation acquired Australian developer Pacific Hydro and its portfolio of renewables projects in 2015, and the company placed an order with Goldwind for the 80MW Punta Sierra project in Coquimbo, northern Chile soon afterwards. And there are many other utility deals in progress that will favour Chinese manufacturers.

In 2016, Goldwind acquired the 160MW Rattlesnake wind farm in the US from RES and then attracted tax equity financing to build the project from MidAmerican Wind Tax Equity Holdings, a subsidiary of Warren Buffet's Berkshire Hathaway Energy, and US lender Citi – the first time a Goldwind project had attracted all-Western financing. Goldwind Americas CEO David Halligan called the deal 'the most significant milestone in Goldwind's history in the US', adding 'our technology is bankable. This kind of makes it official.'

Goldwind is also negotiating to buy the second phase, RES Americas' 160MW Rattlesnake 2 and officials have said in the past that the company has plans to achieve cumulative orders for 2GW of its turbines in the US, Canada and Mexico by 2020 (Davidson 2017). And in April 2017, Goldwind acquired the 530MW project from Origin Energy in Australia. When built, the project, which has a 30-year Power Purchase Agreement (PPA) with Origin, will be Australia's largest wind project. The 149-turbine site located roughly 140km west of Melbourne, will feature Goldwind's new 'smart' 3MW-platform wind turbine (Weston 2017a).

Meanwhile, number two manufacturer Envision is making increasingly bold moves to acquire projects overseas. In October 2015, it bought a controlling stake in a portfolio of 600MW of wind projects from local developer ViveEnergia and formed a 'strategic alliance' with the Mexican company to continue developing the sites, with a view to completing 1.5GW of projects by 2020. Overall, the partnership is targeting 1.5GW on wind projects to be completed by 2020. The Envision Executive Director, Felix Zhang, said that

> Mexico is one of the most promising markets in the Americas for wind power generation in the coming decade, not only a result of the energy reform but also given its untapped wind resources, viable projects and off-takers, as well as the interest of equity sponsors and lenders.
>
> (Weston 2015)

It followed up the deal by winning 90MW in a 2016 Mexican energy tender, securing the rights to construct the Peninsola wind farm by 2018.

In 2016, it took part in the first wind tender under the Macri government in Argentina (see Chapter 4) and was the surprise winner, taking four projects with a total capacity of 185MW out of the 708MW of wind projects awarded, with completion due in 2019. In December 2016, Envision acquired Velocita Energy Developments' French onshore wind portfolio, which includes a 500MW portfolio of projects, mainly in northern France (Weston 2016b).

Overall, then, China is having a huge effect on the global market through becoming the world's biggest single market and through its huge influence in reducing costs throughout the global supply chain. It is also providing momentum for global policy and providing leadership when the US and EU is not in a position to provide that.

As we have seen, China is also boosting investment on a project level throughout the world either through financial institutions, through utilities, or through turbine manufacturers. It will be some time, however since we once again see a Chinese turbine manufacturer – or more than one – at the very top of the wind turbine manufacturers ranking. Chinese OEMs are still relatively reliant on their home markets, and have little interest in making major acquisitions of their competitors.

And as we shall see in Chapter 8, amid all the talk of Chinese competition, the Western companies have not been standing still, but have been moving towards creating global giants prepared to fight it out in the long term.

4 Emerging powers

Another key battleground in the fight to establish wind power as one of the world's dominant energy sources is the big emerging markets. The two most important to date are India and Brazil, both of which have big populations and steadily growing economies. Both countries need to consistently put on huge amounts of new power capacity to keep the lights on – something which in India sometimes appears to be a losing struggle – and both have turned to wind power in a big way over the past decade.

Following closely are a number of countries including South Africa, Turkey, and Mexico all of which have a significant demand for regular increases in generation capacity, and have turned to wind in recent years.

Beyond this leading group, there are a whole series of countries that are power-hungry and have begun to see wind as an alternative to fossil fuels. These include Pakistan, Vietnam, the Philippines and Thailand in Asia; Morocco and Egypt in North Africa; and Argentina, Chile and and Uruguay in Latin America. There are also a group of countries along a windy East African corridor that are also starting to see wind capacity constructed; these include Ethiopia and Kenya.

There are two features of emerging wind-market growth that are of major significance for the industry as a whole. First, in some of the world's potentially biggest power markets wind is already showing itself capable of competing with and beating fossil fuel and nuclear generation on price. Second, given government moves to ensure local content and local advantage, some emerging markets are becoming significant wind-turbine producers in their own right.

I want to look at two of the fastest-growing markets; India and Brazil, both of which have markedly different power mixes and policy frameworks for building their industries.

India

On 30 and 31 July 2012, India suffered the largest power cut in history. It affected over 620 million people or around 9 per cent of the world's population, across 22 states in the north and northeast of the country. Around 32GW of power-generation capacity went offline in the outage.

The blackout was the consequence of a power system that is struggling to maintain coherency and keep up with demand. Smaller, but widespread blackouts are a frequent occurrence in India, along with constant 'brownouts' or lack of tension in the electricity supply system. Forced stoppages in factory production are also common.

Just as serious as the perilous state of India's power system is the fact that 300 million people do not have access to power at all, while power supply to many others is intermittent. Losses at transmission, distribution and consumer level are in excess of 30 per cent. A lack of clean and reliable power means an estimated 800 million people use fuel wood, agricultural waste and livestock dung for cooking and other domestic needs, and the resulting indoor air pollution causes between 300,000 and 400,000 deaths per year, as well as other chronic health issues.

With steady – albeit sluggish – GDP growth, population growth and urbanisation, analysts predict that power demand will hit 300GW by 2020–21, requiring 400GW of generation capacity. This means installing something around 200GW of new generation capacity over the next 6–7 years. And even this, of course, would not be enough to ensure that all of the country's population have access to power.

The country's power-generation mix is extremely problematic. India does not have significant amounts of natural gas, leaving it largely dependent on coal. Its own coal reserves are abundant, but of low fuel quality and high ash content. They are also mainly buried under protected tribal and forest areas, and mining projects create fierce local opposition. As a result, India faces a severe coal shortage. Coal imports were a record 138 million metric tons in the 2012/13 fiscal year (Williams and Menon 2013), and India is now the third-largest coal importer in the world after China and Japan. Meanwhile the cost of imported coal is likely to rise to around US$25bn by 2016/17 from US$14bn in 2012/13, frustrating Indian government attempts to reduce its current account deficit.

With its hopelessly inadequate ports and rail infrastructure, the country is literally choking with the stuff. Many power plants typically have reserve coal supplies to last no more than a single day of operations. Coal is damaging both to the quality of life of India's citizens, and to the country's international reputation. By 2012, it had helped to make the country the world's fourth-biggest carbon emitter, and one of the fastest growing, with some analysts predicting that the country's emissions will be on a par with China – if not higher – by 2020. Compared to the average of thermal plants in the EU countries, India's thermal plants emit 50 per cent to 120 per cent more CO_2 per kWh produced.

India started to install its first wind projects in 2000, and the amount of MW installed annually grew to just over 3GW in 2011, before falling back temporarily to 2.34GW in 2012. The cumulative installed capacity at the end of 2012 was 18.4GW, according to GWEC. The sub-group for wind-power development appointed by India's Ministry of New and Renewable Energy to develop plans for the 12th Plan period (April 2012 to March 2017) fixed a reference target of 15,000MW in new capacity additions, and an aspirational target of 25,000MW.

Analysts point out that in certain conditions, Indian wind projects have a lower levelised cost of energy than new coal plants. Wind projects are also quicker to develop – they can often be realised in 2 years from initiation to completion, compared to 5–7 years for a new coal plant project. Wind is also scalable and projects can be built with a relatively small capital outlay and then expanded. The combination of these two factors makes wind a powerful tool to solve India's power shortage and stop its carbon emissions from growing.

The main government incentive throughout the early period of India's wind growth has been Accelerated Depreciation (AD), which allowed companies to claim a depreciation benefit of up to 80 per cent of the cost of equipment in the financial year after project commissioning. Although AD was successful in incentivising the first stage of India's development, it created a number of anomalies in the wind market.

With AD, India did not create wind-power developers as in most markets. With a few exceptions, the companies investing in wind projects were so-called 'tax investors'. These were non-energy companies, with small – often 1–3 turbine – projects. The tax investors had no expertise or willingness to get involved in carrying out the range of development and engineering tasks that a wind farm involves. So the other result has been that wind-turbine manufacturers in India have traditionally been developers. This means acquiring the land for the project, which as we will see is a crucial skill with sometimes less than salubrious aspects to it. It also means carrying out a full 'turn-key' contract, providing what EPC services may be necessary, including building access roads, installing the turbines, and building substations and transmission lines if needed.

Analysts have long argued that AD has not helped in creating the scale of projects that is needed to really push down wind-energy prices, while the financial nature of the investment means that investors are not focused on achieving cost-of-energy improvements. It has also meant relatively higher working capital requirements for turbine manufacturers, compared to a supply-only or supply-and-install model that is common in many projects outside India.

Over the most recent period, the Indian wind-power market has been transformed as new frameworks that reward power production rather than capital expenditure have emerged. These included federal generation-based incentive (GBI), as well as state feed-in tariffs and renewable portfolio/green certificate support schemes. As we shall see, the new structures and the phasing out of AD have favoured a new type of large-scale developer whose interest is making profits from wind-power generation. It has also shaken up the market among wind-turbine suppliers.

A new type of market

Indian wind developer Mytrah had many sceptics when it started operations in 2010. Its aim was to set up as an independent, privately financed energy producer that in a short time would become one of India's largest wind-power operators.

Starting as Caparo Energy and carrying out an initial placing of shares on London's AIM stock market, Mytrah has relied on successive rounds of mezzanine finance to carry out its plans, which involved reaching about 1.5GW in operational wind assets in 2015.

It quickly showed the size of its ambition by signing turbine deals for 3GW and 2GW with Suzlon and Gamesa, respectively. It has also managed to steadily build out its portfolio, as well as acquiring other assets. It had around 310MW of wind farms in operation at the end of June 2013, and was expecting to increase this to 548MW by the end of the year. Despite capital spending and paying back loans, the company made a profit of US$3.9m during the first six months of 2013, up from US$2.6m during the same period in 2012.

Chief Executive Ravi Kailas told *Recharge* 'Our numbers seem large, but in relationship to the overall market it is small', citing the total 65GW capacity expected to be installed in India by 2020. 'What we are doing in five to six years' period is do-able. It is a challenge, has not been done before but is not impossible' (Backwell 2011b).

Mytrah is indicative of a new generation of independent power producers set on profiting from Indian wind-power generation. But the expansion of the market has not been smooth.

While AD has been phased out, the initial generation-based incentive (GBI) meant to replace it was widely seen as having been set at an inadequate level and was suspended in April 2012. The government reintroduced it at a higher level in August 2013, but not before the expansion of the Indian wind market had suffered a serious hiatus, with installations in 2012 at 1.7GW, 47 per cent lower than in 2011. Another key area of negotiation are state feed-in tariffs, which are set by the regulatory commissions of each state, and which are not adequate to stimulate wind power in some states. The green-certificate market, which is linked to state renewable portfolio standards, has great promise, but is currently not effective because there are insufficient penalties to hold states responsible for not meeting standards, and utilities are not bound to buy clean power.

The rise of Suzlon

Nestled amid a growing cluster of skyscrapers in the traditionally genteel and academically orientated, but now fast-sprawling city of Pune lies the One Earth Centre, the headquarters of India wind-turbine group Suzlon. The 100-per-cent-renewables-powered centre, designed by US-born architect Christopher Charles Benninger, is spread across four hectares and can house 2,300 people.

Indian businessman Tulsi Tanti spotted the potential of wind relatively early and set out to create a global group – but one based in an emerging market – that would bring cheap wind power to the masses. Tanti captured the zeitgeist, with his huge optimism and energy and high-profile engagement with global climate forums and green business groups, becoming one of India's richest men (for a while at least) in the process.

Tanti was managing his family's 20-employee textile company in the early 1990s and said he became interested in energy as he experienced the effects of India's shaky power grid and realised that electricity was his firm's second-highest cost after raw materials. He set up Suzlon in 1995 – entering into a technical collaboration agreement with German company Sudwind GmbH – and adopted a business model where customers would put up 25 per cent of the cost of a wind farm and Suzlon would arrange bank finance for the other 75 per cent. After overcoming initial reluctance, by 2008 over 40 Indian banks were financing wind-power projects for Suzlon clients.

Tanti sold the textile business in 2001 and by the 2005/06 fiscal year, Suzlon was supplying over half of India's wind market. Expansion into the US – Suzlon achieved its first sale there in 2003 – China, Australia, Brazil and other markets was taking place simultaneously. Turbine installations passed 1GW in the 2004/05 fiscal year, and by 2012/13 had crossed the 20GW mark.

Figure 4.1 Suzlon's founder and Chairman, Tulsi Tanti (Source: Christopher Bangert/Danish Wind Industry Association).

In 2006, Suzlon paid US$565m to acquire Belgium-based gearbox manufacturer Hansen Transmissions, as it deepened its holdings in the supply chain. In 2007, Tanti engaged in his most ambitious move yet – the purchase of a majority stake in German turbine manufacturer and offshore specialist REpower – which valued the company at US$1.6bn. The battle to win majority control over REpower put Suzlon head to head with French nuclear giant Areva, which was planning its own entry into the offshore wind market, in what newspapers dubbed 'France vs. India for a piece of Germany'.

Tanti won the battle for majority control after five months of intense rival bidding with Areva for REpower shares, and through gaining the support of Portuguese engineering and renewables company Martifer, which had a 23 per cent stake in REpower. The two companies bid 20 per cent over the price offered by Areva in a public offer for shares through a Suzlon-financed special-purpose vehicle, with an agreement that Suzlon would buy Martifer's entire REpower stake two years later. By 2009, Suzlon controlled 90 per cent of REpower, and by October 2011, it had completed a so-called 'squeeze out' process to acquire the share of remaining minority shareholders. The move catapulted Suzlon into the top four global manufacturers, but it was also the major contributor to saddling Suzlon with over US$2bn of net debt.

Tanti's business model was always based on highly optimistic forecasts of global wind-power growth. These were not unrealistic, when the wind industry was clocking up spectacular annual installation figures in the mid/second half of the decade. As we shall discuss in Chapter 6, however, the impact of the global sub-prime crisis on financing, the failure of the Copenhagen climate talks and the overhang in wind-turbine manufacturing capacity, along with a major issue with cracks to rotor blades affecting its turbines between 2007 and 2009, were soon to take a heavy toll on Suzlon's business.

Added to the international factors came a slowdown in the Indian market, as the government moved to phase out AD. Newer incentive schemes such as the GBI or green certificates were either not set at adequate levels or had not been fully implemented. Suzlon's debt problems meant that by 2012 it was facing a major squeeze on its working capital that left it unable to fully execute its extensive project pipeline, and it lost its place as India's number one OEM in terms of annual installations – although it was still by far the biggest in cumulative terms – to Wind World (India), the former subsidiary of Enercon in India. We shall look at the fortunes of Suzlon, and its chances of bouncing back, in Chapter 7.

Not for the faint-hearted – foreign OEMs in India

India's market is not for the faint-hearted. Obtaining land and building permits means negotiating India's state bureaucracy, and this requires a particular local set of skills and experience.

As in many markets, Vestas was a front-runner in India, and was the second-largest supplier in the Indian market, operating through a joint-venture subsidiary with local partner RRB. Fellow Danish company NEG Micon was in first place through NEPC Micon, which was a joint venture with the NEPC group. Like other OEMs in India, both companies engaged heavily in wind-farm development including land procurement.

When Vestas completed its Micon acquisition in 1996, NEPC Micon was split up. Vestas also split with RRB in 2004, as the company consolidated the Micon operations, and moved to a turbine supply-and-installation-only model. It quickly found that it could not maintain market share with this strategy and by the end of 2011 orders had dried up. In September 2012, Vestas announced it was scaling down its India operations, 'to focus on providing value-added service and maintenance to existing Indian wind-power plants' (Ramesh 2012). As we shall see, key staff left in frustration to join rivals.

Now that India's wind market is moving to a new model, with development being carried out by large independent power producers (IPPs), Vestas may have a chance to re-enter the market. But with competition from Suzlon and several other highly aggressive players, it won't be easy.

Enercon India goes 'rogue'

As in China, there are also serious issues with the protection of intellectual property rights as can be seen in the case of German turbine manufacturer Enercon, which found itself fighting a losing battle against a 'rogue' subsidiary.

Enercon made a good start in India, setting up Enercon India (EIL) in 1995 as a majority-owned joint venture. By 2010, it had established itself as the second-largest supplier in India after Suzlon, with its highly regarded 800kW direct-drive turbines.

However, in 2007 a dispute between Enercon and its local partners, the Mehra Group, flared into the open after two years of worsening relations. Even though Enercon held 56 per cent of EIL, the company was in effect controlled by the managing director, Yogesh Mehra, who managed to exclude the parent company's representatives from the board.

A legal case brought by Enercon in 2007, accusing its Indian partners of concealing the company's state of affairs and financial mismanagement, resulted in a court ruling that preserved the Mehra family's control of EIL. From 2007, there was no further cooperation between the two companies. Enercon no longer vouched for the 800kW turbine EIL continued to build and sell. The Indian company continues to use the parent company's logo and literature on its website, and continued to do well, signing major deals, particularly with Hong Kong-based developer CLP Group for projects like the 113MW Andhra Lake wind farm.

In early 2011, EIL managed to convince a patent court in Chennai to annul 12 of Enercon's patents, arguing that the patents did not represent any genuine innovation. 'In view of this analysis and the finding herein, it is very clear there is no inventive step in the invention claimed in the impugned patent and the invention claimed is obvious to any skilled person in the art', reads one of the rulings seen by *Recharge* (quoted in Backwell 2011a).

Enercon argued that the rulings had major implications for other turbine manufacturers in India. Enercon lawyer Stefan Knottnerus-Meyer says the court based its decision in part on the premise that India's national interest in developing renewable energies has priority over a company's patent rights. 'In keeping with this argument, almost every new patent could be nullified in the future, in the name of India's development interests', he adds (quoted in Backwell 2011a).

The rulings left EIL free to carry on producing its turbines and prevented the German group from making any exclusive deal to license the turbine models to another Indian group. On the other hand, it remains to be seen if EIL is capable of designing and building new and bigger turbine models, which are likely to represent an increasing part of demand in India, as the market moves to a new wind-power development model. For now, however, EIL, which is now called Wind World (India) continues to do nicely; in 2016 it was the second biggest turbine company in India in terms of cumulative installed capacity.

Gamesa moves in

As we have seen, Vestas' fortunes in India declined when the company moved to a turbine supply-and-installation-only model. Vestas' loss to a great extent turned into a gain for Spanish turbine manufacturer Gamesa, when India boss Ramesh Kymal became the new Gamesa India's CEO, taking key members of the team that had built up NEPC Micon/Vestas with him.

After entering the market in 2010, Gamesa became the third-largest supplier in India in 2011 with a market share of 10 per cent. It built a series of new factories including two nacelle assembly plants in Chennai and blade and tower plants in Baroda, in Gujarat state.

Kymal says that he had been opposed to Vestas' decision to pull out of project development, as well as what he sees as a failure to understand local realities. He says that when he started with Gamesa he was able to persuade the company's global management to start in India with the extremely grid-robust Gamesa G58 850kV turbines to build its brand, rather than a newer, but more sensitive model.

Development activities, which Gamesa carries out worldwide, and the ability of Kymal's team to gain access to land have been key to the company's rapid expansion in the country. Gamesa is also benefiting from the shift in the market from traditional 'tax investors' to IPP clients and has chalked up some key successes for its G58 and 2MW G97 turbines. Kymal told *Recharge* in Chennai in 2011 that the company went from 100 per cent traditional clients in 2010, to supplying 30 per cent to IPPs this year, and this was expected to increase to 60 per cent in 2012.

In May 2011, Gamesa pulled off a major coup by signing a US$2bn, 2GW supply deal with developer Caparo Energy India (now Mytrah Energy). Under the terms of the deal, Gamesa will provide Caparo with G58 850kW and G97 2MW turbines between 2012 and 2016.

As well as Gamesa, there are also several other companies which are aggressively growing in the market, as well as more well-known players such as GE. One is Indian company ReGen Powertech, which has a licensing agreement with Goldwind-controlled turbine designer Vensys for a highly regarded direct-drive turbine. The company grew fast in 2010–11 and was in third place by new installations in 2012, behind Suzlon and Wind World (India). Managing director Madhusudan Khemka says nearly all ReGen Powertech's sales come from IPPs, rather than smaller 'tax investors'.

Where next?

As we have seen, the phasing out of AD and the suspension of the GBI in April 2012, along with other factors such as worsened financing conditions, meant that the promise of a speeding-up of India's wind development was postponed in 2012 and 2013. But it will come, as India's fossil-fuel-based energy system becomes increasingly problematic.

Gamesa's Ramesh Kymal, who also chairs the Confederation of Indian Industry's renewable-energy council, points out that wind could already compete with conventional power plants if fossil-fuel subsidies were removed. 'This whole change is coming', he says. 'That is why IPPs are becoming much more viable, because the FIT will go up, REC prices are already going up and we are lobbying hard to have the GBI extended and increased. I am looking at a huge market of at least 5–8GW per year of steady growth in India' (Backwell 2011c).

There are huge challenges, such as implementing central government policies through the states and through local bureaucracy. The physical challenges of constructing wind in India – the lack of decent roads along which to transport wind-turbine blades and towers, and the chronic lack of transmission lines in many areas, which can have crippling effects on both planned and operational projects – should not be underestimated, but they can be overcome.

Since the first edition of this book, the Indian wind market has once against sprung into a faster growth mode, despite reductions in tax incentives and other support mechanisms. India's market is set to continue to grow at a faster rate as renewed growth is being increasingly driven by price competition, either through auctions or through corporate power purchase agreements (PPAs) and as the government moves steadily towards fulfilling its Paris Agreement (COP21) commitments to raise its share of non-fossil power generation to 40 per cent of the total, and provide electricity to the 240 million people in India who remain without access to electricity. In 2015, the government set a new short-term wind-power target of 60GW by 2022, and the industry is well on track to meet it, having reached installed capacity of 31GW in March 2017.

In 2016, the industry broke all previous records and installed 3.6GW of capacity – putting it in fourth place globally, both in terms of annual and cumulative installations, while in the Indian fiscal year 2016–17 installations were 5.4GW. In order to meet the 60GW target, the industry will have to continue to install over 5GW over the coming years.

Meanwhile the turbine market continues to be vibrant, with a resurgent Suzlon and a number of established and up-and-coming challengers. According to GWEC and the IWTMA, the top-five OEMs in terms of cumulative installed capacity in India are Suzlon (35.4%), WindWorld (18%), Gamesa (10.1%), Vestas (7.6%), Regen (7.3%) and Inox (5.68%). A number of companies have made important new investments over the last couple of years. LM Wind Power set up its second blade factory in Vadodra, Gujarat. Senvion (formerly Suzlon controlled REpower) started up its operations in India and acquired the Kenersys manufacturing facility. Gamesa set up a new factory at Nellore in Andhra Pradesh; Vestas opened a new blade manufacturing plant in Gujarat; Spain's Acciona entered the market in 2016, while Chinese companies Envision and Sany Global were also planning to enter the market. The current manufacturing capacity in the country is around 10 GW.

Where is Government policy going?

While India's wind industry is on the right path there is significant regulatory uncertainty ahead.

The Renewable Purchase Obligation (RPO) and renewable energy certificate linked to it, which we discussed above has not been considered to have been a great success, despite some 28 states and union territories adopting RPOs. This is largely due to the non-compliance and weak enforcement of the RPO by the states and market regulators.

Meanwhile the government, in its latest budget announcement for FY 2016/17 reduced the Accelerated Depreciation tax break from 80 per cent to 40 per cent. And the Generation Based Incentive could come to an end at the end of the 2016/17 financial year.

While these two factors led to a rush to complete projects and thus contributed to the strong rate of installations at the end of the 2016/17 fiscal year, growth in the future is set to come from large-scale power auctions. The first auction for 1GW of capacity was held successfully in early 2017, with winning bids coming in at INR 3.46 (USD 0.052/EUR 0.049)/kWh. Another 4–5GW tender was expected to be tendered by March 2018. The move to auctioning has been accompanied by a more realistic and proactive approach to India's wind sector by the government, which is now facilitating the development of a national wind power market

> We do not apply interstate transmission charges. The benefit of that can be reaped by states like Uttar Pradesh and Bihar, which do not have sufficient wind capacity. These states can utilise the power from states which have a higher wind capacity at lower rates and can also fulfil their renewable purchase obligations (RPO).
>
> (Tenddulkar, 2017)

To address grid integration challenges, the Indian government has initiated the

"Green Corridor" programme, which aims to increase connectivity between India's regional grids with its national grid. The programme will facilitate the transfer of power from the high renewable energy installation states to other parts of the country, through a series of new transmission lines, HVDC terminals and other equipment. Other policies that have been introduced include a draft framework to facilitate hybrid wind-solar projects and a number of other helpful guidelines.

To keep growing at the required rate, India will need to enforce RPOs, and create a true interstate power market with low or no transmission fees. More sophisticated grid integration of renewables will be needed and wind operators will have to become better at scheduling and forecasting as well as starting to participate in balancing and ancillary services markets (GWEC and IWTMA 2016).

GWEC predicts that if the current pace of installations were sustained over the medium term, Indian wind-power installations would be on a path to crossing the 300GW mark around 2040, and provide the much-needed cheap power for meeting India's development and environment agenda.

Brazil

Brazil takes off

Brazil shares India's high growth potential, with steady expansion in economic outcome and power demand forecasted, boosted in the short term by the expectation of a 'Brazilian decade' with highlights including the country's hosting of the 2014 football World Cup and the 2016 Olympics.

Like India, Brazil faces a struggle to keep the lights switched on, although its power matrix could not be more different. Around 80 per cent of Brazil's power comes from hydroelectric power, while thermoelectric power from gas, fuel oil and coal makes up just over 16 per cent. Nuclear power makes up 2 per cent of the total supply.

Brazil's government sees developing clean energy as essential to continuing to project itself as a global leader in terms of 'soft power' as it pushes for a place at the top table of world politics (see Backwell 2013b). Traditionally the availability of hydrocarbons has been limited, and this allowed Brazil to become a global pioneer in biofuels in the 1970s. The arrival of a boom in 'pre-salt' ultra-deep-water oil exploration has not brought any immediate large increase in flows of natural gas to the mainland, and it is still debatable how big such a flow will be in the future. Hydroelectric power will remain the mainstay of Brazil's power system. While this is positive in terms of carbon emissions, it is difficult to add large amounts of new power, given that hydro projects require huge capital investments, long-term planning and have major impacts on the environment and local populations – big projects such as Belo Monte have caused huge public controversy. The hydropower supply can be sharply curtailed if there are several consecutive drought years, as there were in the run up to the 2001–02 energy crisis, which Brazil overcame without blackouts only by reducing consumption by 20 per cent over an eight-month period. Hydro-generation relies on highly seasonal patterns of rainfall, and this can bring Brazil's power system close to the brink.

Government policymakers who were once sceptical about wind power have discovered that wind power is highly complementary to hydropower, with Brazil's winds blowing most strongly during the winter dry period. As a report by GWEC in 2012 says:

> Hydro and wind power are perfect partners in Brazil. Not only are the country's windiest areas located conveniently close to demand centres, but in addition, the variable nature of wind power is best accommodated in a highly flexible system such as one dominated by hydropower. Furthermore, wind power can help alleviate some serious energy security concerns in Brazil, especially during the dry winters.
>
> (GWEC 2012)

One of the consequences of the 2001–02 energy crisis was the creation by the new PT administration of an auctioning system in 2004, as part of a series of reforms aimed at ensuring that adequate amounts of new power would be brought online. As part of the reforms, the government also created a new company, EPE, tasked with long-term energy planning and deciding – among other things – how the auctions should be designed and which energy sources should be included, based on the criteria of energy supply security and cost-effectiveness.

Brazil has extremely good wind resources. In large parts of the northeast of the country, the wind is constant – industry officials compared it to a vast warm hair dryer – and this has allowed Brazilian wind projects to clock up some of the highest average capacity factors in the world, at in excess of 50 per cent. In addition, wind capacity can be quickly brought online, as industry officials explained during a series of meetings in the early part of the decade, as they struggled to get wind taken seriously in Brazil.

The government established the legislation for the first support system for wind called PROINFA during the administration of Fernando Henrique Cardoso, although this was not implemented until the first Lula government by the then Energy Minister – now President – Dilma Rousseff. Government officials continued to be fairly sceptical, however, and the breakthrough moment did not come until spring of 2009, when wind-industry bodies organised a trip to the then booming wind market of Spain for key Brazilian politicians and government officials, including members of the renewable-energy committees in both the upper and lower houses of the Brazilian Congress. 'Seeing the scale of the industry, and most importantly the Red Electrica control room, from which the whole Spanish power system is controlled, caused cascading epiphanies in the minds of Brazilian officials and politicians', says GWEC's Steve Sawyer (personal communication).

Among those on the trip was EPE President Maurício Tolmasquim who organised the first capacity auction later in 2009, and who says that 'wind's moment' has arrived. 'I was never against wind power', Tolmasquim told *Recharge* correspondent Milton Leal in his offices of the 11th floor of an imposing commercial building overlooking Rio de Janeiro's beautiful Guanabara Bay.

> Inside the Energy and Mines Ministry, I was one of the most favourable. During the time of [the alternative-energy incentive scheme] PROINFA, I defended the need to give a chance for wind. But at the time price was a barrier … The planner can't have a passion for, or hate a form of energy. He has to have a rational view. Above a certain price level, I couldn't justify contracting wind.
>
> (Leal 2013a)

Today this is possible, as Brazilian wind projects have been bidding for capacity at prices that are arguably the world's lowest and beating fossil-fuel projects on equal terms along the way.

At the December 2012 A-5 auction ('5' being the time period in years within which the new power needs to come online) Brazil awarded contracts to four developers to sell 281.9MW of capacity from ten wind farms due to be completed in January 2017. The average rate was R$87.94/MWh (US$42.2/MWh), with the lowest prices being R$87.77/MWh. Average prices were 12 per cent cheaper than the R$99.58/MWh average price in August 2011, which at the time was considered by some to be the lowest price for wind power in the world. These projects compare with the cheapest power purchase agreements (PPAs) in the US, which come in the range of the mid-US$50s per MWh if the government's production tax credit is factored in.

Brazil's highly competitive auction system has created a number of dynamic local developers, whose business models are extremely aggressive. The most impressive is Renova. Run by 41-year-old former McKinsey consultant, Mathias Becker, Renova has expanded at breakneck pace. When Becker was appointed at the start of 2012, Renova had 110 employees and zero wind installations. By September 2013, it had 294.4MW installed (although not grid-connected), 1.29GW of PPAs, a 12GW pipeline across Brazil, a 280-strong workforce and an average annual growth rate of 30 per cent mapped out until 2017 (based on PPAs signed before August 2013).

Becker became involved in Renova when one of his consulting clients, state-owned utility Cemig, was looking for a way to enter the renewables sector. Becker suggested investing in Renova – which was founded by Ricardo Delneri and Renato Amaral, and which at that time had won 462MW in national tenders. Cemig bought a 25.8 per cent stake in the company for R$360m (US$158m) through its subsidiary Light. In August 2013, Cemig announced that it was pouring R$1.41bn (US$ 0.64bn) into the developer via another subsidiary focused on generation and transmission, joining Light and a third stockholder, RR Participações, in a block controlling at least 51 per cent of Renova's shares. Six months into Becker's reign, the Brazilian national development bank, BNDES, bought a 12.2 per cent stake in the company for R$260.7m (US$117m).

Having both BNDES and Cemig on board has paid dividends for Renova, significantly reducing the cost of its financing. The stock market has also shown more confidence in the company – Renova's share price rose 200 per cent after the initial public offering in 2010.

The force driving Renova's growth is the extraordinary winds of north-eastern Brazil. At Renova's Alto Sertão complex in Bahia, Brazil, the winds blow in from the east at an average eight metres per second, with a daily lull from around 11am to 4pm that makes it safe enough to fly a helicopter – or install a turbine.

Capacity factors are a spectacular 45–50 per cent and there is a proven resource of 6GW across the 11,000sq km that Renova has leased from local landowners. An excellent internal rate of return on the 294.4MW first phase is set to rise for the second and third phases. It is little wonder that the company refers to Alto Sertão as its 'gold mine'. 'We want to reproduce the same Gold Mine effect [in other parts of Brazil], having Gold Mine 1, 2, 3, 4, etc.', Becker told *Recharge* (Leal 2013b).

But not everything went Renova's way. Its wind farm projects suffered long delays in starting up because the construction of necessary transmission lines and other infrastructure by state-owned utility Chesf fell behind schedule. Renova was being paid by the state for the electricity it would have produced, but this was hardly of benefit to Brazil's consumers and the company had to pay for oil-fired generators to spin the turbines' rotors for maintenance purposes.

Sérgio Marques's big bet

An aggressive development can sometimes spill over into what some analysts consider as 'reckless', as in the case of rival developer Bioenergy, which is the creation of controversial Brazilian businessman, Sérgio Marques.

Marques – like Becker, relatively young in his late 30s – has almost single-handedly grown his company from a one-man band to a multi-million-dollar business using little more than his own creativity, drive and powers of persuasion. His projects have been the lowest of the lowest priced in Brazilian wind auctions, with 201MW of Bioenergy projects bid at the record R$87.77/MWh (US$40/MWh) price in the December 2012 A-5 tender.

In 2002, Marques was an executive at a subsidiary of ABB, which was then trying to break into the Brazilian wind sector. When the Swiss–Swedish technology giant pulled the plug on the unit, Marques set up Bioenergy and took all the projects he had been developing for ABB with him.

Between 2005 and 2009, Bioenergy signed about 400MW of long-term PPAs, either through national tenders or on the open market. The money he gained from selling these PPAs was used to fund the company's only two operational wind farms, the 14.4MW Aratuá 1 and the 14.4MW Miassaba 2 – the latter being the first in the country to deliver power to the unregulated market. By late 2013, Bioenergy had signed around 800MW of PPAs.

Marques is able to make extremely aggressive bids for capacity because the success of his projects are predicated on them being built ahead of time. For the

R$2.5bn (US$1.1bn) Paulino Neves wind project, for example, Marques bet that the R$1.4bn (US$0.6bn), 351.9MW first phase of the 640.9MW complex could be built ahead of time, so that the turbines can reap huge amounts of money on the unregulated market before part of the PPAs signed in the regulated market start in March 2014. Marques was so confident that he began to sell his pre-PPA output in the market. He estimated that Bioenergy could see a R$300m (US$135m) profit in 2014 and millions more in 2015, at the then spot price or around R$120/MWh (US$54/MWh) in 2014, and in fact was expecting prices to be closer to R$170 (US$77) – which would yield far bigger profits.

However, if Marques' projects were to fall behind schedule, Marques' company faced ruin, as the penalties for those who sell but don't deliver energy in Brazil's system can be astronomical – companies are obliged to buy the power it has contracted to supply at spot prices, which can fluctuate wildly and reach peaks far above the average price.

In the event, Marques' business model proved to be far too aggressive, as it became clear that there was little possibility of being able to build the projects in Maranhão – which at the time had zero existing wind capacity – in the timeframes foreseen by Marques.

In June 2015, Bioenergy applied to the regulator Aneel to revoke all of its supply contracts for 19 planned wind farms in Maranhão and Rio Grande do Norte, for a total of 547MW, in the midst of a legal dispute with state energy planning company EPE. No work had begun on any of the plants, which had delivery dates for 2013, 2014, 2016 and 2017. 'We are not going to implement the wind farms,' said Marques, 'as the fines for delay are higher than the profitability of the projects. It's become unviable, something disproportional.' (see Costa da Paula 2015).

The OEMs pile in

The creation of a dynamic Brazilian market, combined with demanding local content conditions, has been extremely successful in drawing in a large number of international turbine suppliers.

The local content rules operate through the national development bank, BNDES, which is practically the sole local source of financing for wind-power developers and offers extremely attractive – sometimes negative if inflation is taken into account – interest rates. In order to qualify for financing for bids in Brazil's power tenders, developers' designated turbine suppliers need to meet the BNDES conditions. These conditions in turn have become more complex, moving from an overall percentage of all the turbine equipment used in a project, to a 'through the turbine' approach that requires manufacturers to meet different percentages in several different areas of the turbine.

The first movers in Brazil were Germany's Enercon through its Wobben Windpower subsidiary, Argentina's IMPSA, Suzlon, GE, Alstom and Vestas. Wobben's turbines are highly admired in the Brazilian market, while IMPSA has pursued an aggressive growth model based on turbine sales and participation of its own development arm, Energimp, in the power tenders. IMPSA had all its manufacturing based in Brazil from the start, giving it a significant advantage in meeting the second phase of local content rules. The company had an impressive 1.26GW of turbines installed and on order in Brazil; the country's largest wind-turbine production capacity (1.4GW) and formidable political contacts within federal and regional governments.

IMPSA Vice-President José Luis Menghini said the company had poured US$250m into its Brazilian infrastructure so far, with a further US$60m being spent on a new 400MW nacelle factory in the southernmost state of Rio Grande do Sul, which would also supply the Argentine and Uruguayan markets. IMPSA's existing 1GW nacelle plant at the Port of Suape, in the northeast state of Pernambuco, is operating at full throttle, as the company has to deliver most of a 900MW-plus backlog by the end of next year.

From 2008–2014, IMPSA won 1.26GW of orders in Brazil – a 15 per cent slice of the market – but most of that (803MW) was ordered by Energimp. The remaining 460.5MW was split between state-owned generators Chesf (180MW) and Eletrosul (234MW); Tecneira, a developer of mainly small-scale projects, owned by Spain's ACS group (42MW); and Australia's Pacific Hydro (4.5MW).

However, from 2014, IMPSA's fortunes took a sharp turn for the worse. The company experienced a number of problems with its direct-drive machines – based on a well-regarded design licensed from Goldwind-owned Vensys – and components such as blades and towers, and a dispute with Brazilian state electricity holding Electrobras. Meanwhile, its Argentine parent company was suffering wider problems due to late payments from Venezuela. In late 2014, IMPSA defaulted on its foreign debt and was forced to put up its two Brazilian factories and other operations up for sale. While the parent company has come back from the brink and made a deal with its creditors, no buyer has so far come forward – although persistent rumours suggest Goldwind was interested – and the Brazil factories remained shut as of September 2017."

Alstom wins big with Renova

The size of the opportunity that Brazil represents for manufacturers became clear in February 2013, when Alstom signed a €1bn (US$ 1.4bn) deal to supply at least 1.2GW to Renova. Alstom described the deal as its 'most important' ever in onshore wind. It has built a second factory in the south of Brazil, to complement its plant in Camaçari, Bahia, where it has introduced a double shift, increasing capacity to 600MW annually.

Alstom's tie-up with Renova capped a run of contracts won since mid-2010, when the manufacturer was tapped by São Paulo-based Desenvix to supply 57 of its 1.67MW ECO86s for the Brotas development in Bahia. In 2011, Alstom signed a €200m (US$275m) agreement to supply and maintain 41 ECO86 machines for a trio of Brasventos wind farms in Rio Grande do Norte state. In 2012, Alstom signed deals with Odebrecht Energia – a 108MW order to kit out four wind farms in Rio Grande do Sul with 40 ECO 122s; Casa dos Ventos – a

€230m (US$316m) deal to supply 68 ECO122s in Rio Grande do Norte; and a pair of contracts with infrastructure group Queiroz Galvão for Eco122s for two complexes in the same state. 'Latin America, especially Brazil, will be one of the centres of wind development in the coming years', says Alstom's Brazil country manager Marcos Costa. 'Brazil will also be the centre of development and production for wind farms in other Latin American countries' (Backwell 2013c).

The Renova deal is particularly notable in that the Brazilian developer had been working previously with rival OEM, GE. Officials from Alstom say the key to winning the contract was offering a range of turbine options that could be tailored to the wind conditions of each individual site. Alstom and Renova have set up a joint committee, and are working together to optimise Alstom's turbines for the idiosyncratic winds of the Caetité region. 'We have bet on the best business plan, not on the cheapest machine', says Becker (Leal 2013b).

Because of the Alstom deal, the manufacturer was able to convince its blade supplier, Tecsis, and tower maker Torrebras to build factories in Bahia – dramatically reducing transportation costs for Renova. Blades for the GE turbines currently take 25 days to travel the 1,500km from São Paulo state. For the third phase, the blades and towers will travel less than 700km. 'Renova is almost a market in itself for us', says Marcos Costa, Alstom's Brazil country manager. 'Brazil is the bigger market and within it, Renova is a super market' (Leal 2013b).

Gamesa bets on Brazil for growth

As we will see in Chapter 7, Spain's Gamesa spent the early part of the current decade struggling to reinvent itself, after orders from the Spanish market dried up, and sales to its shareholder and partner Iberdrola were no longer enough to guarantee growth.

In a move to capture sales from fast-growing emerging markets, Gamesa made an aggressive entry into India's wind market in 2010–11, and was simultaneously making moves in Latin America, an area in which many Spanish companies have traditionally seen themselves as having an advantage.

In Brazil, Gamesa set up a manufacturing plant in Bahia state, and was a big winner in the new power tenders, winning over 600MW of orders between 2009 and 2011, using a strategy of competing with ultra-low margins to gain critical mass in the market place.

By the end of the third quarter of 2013, Latin America was the destination for fully 51 per cent of total sales, with the bulk of this coming from Brazil. Gamesa was one of the first companies to confirm that it had complied with the latest BNDES local content rules and obtained FINAME approval.

In the August 2013 reserve auction, companies that had registered to use Gamesa turbines won at least 462MW of capacity, making it the single biggest winner, and the company announced plans to expand its Bahia plant to increase output.

How many companies does Brazil need?

Brazil's combination of competitive auctions and local content rules through the BNDES can be considered a success in terms of fostering the fast growth of the wind sector at competitive prices and building up a new manufacturing industry. But the policy has perhaps been too successful. At a time when growth opportunities in the global wind-power market were relatively scarce – as we have seen, Western companies have been largely shut out of China, the most dynamic market in the period – OEMs have piled into Brazil.

As of early 2013, there were at least ten companies manufacturing or planning to manufacture in Brazil: Enercon/Wobben, IMPSA, Vestas, Gamesa, Alstom, GE, Acciona, WEG, Siemens and Suzlon. For a market that is likely to be around 2.5GW per year – in a fairly optimistic estimate – this is clearly too many to sustain. Competition and pressure on margins have been intense.

Gamesa's country manager Edgard Corrochano said, 'With a 2GW-a-year market, the most logical thing would be for three to four – most likely three – manufacturers to stay in the market, and we will be one of them.' To make the grade, he believes 'You need decent volumes, and we have that, along with maybe two more' (Backwell 2013d).

The new FINAME regulations will force companies to make significant new investments in manufacturing, and companies are making hard calculations whether to simply fulfil existing orders or stay the course and invest to be able to take part in future tenders. 'Meeting [the requirements] is hard work, but we have been working on it continually for three years', Corrochano says (Backwell 2013d). For potential competitors, 'It's not easy to come in and do now what I have already done', he says, adding that those companies that want to stay in Brazil 'are going to have to make big investments' (ibid.).

As of late 2013, Wobben, IMPSA, Alstom, GE and Gamesa said they had complied with the new regulations, while Acciona said it was on course to do so. Vestas, one of the leaders in terms of constructing wind farms following success in the early tenders, announced in June 2014 that it would invest R$100m ($43.5m) to ramp up manufacturing capability in Brazil and meet the FINAME requirements.

Can Brazil's wind power stay cheap?

FINAME 2 is likely to have the effect of driving consolidation in the Brazilian turbine market. But it is also driving up turbine prices and threatens to undermine the price advantage that the Brazilian wind sector has enjoyed until now.

In 2013, turbine prices increased by around 20–25 per cent, because of the stricter local-content rules, helped by a good level of power demand at the latest tenders. Several turbine companies are having to reassess their involvement in the market given the level of competition and renewed pressure on margins, while there has also been a lot of grumbling over the onerous level of inspections and declarations required by the new rules, as well as the difficulties of obtaining

Figure 4.2 Alstom's Senior Vice-President for Wind, Alfonso Faubel (Source: EWEA).

some components locally at realistic prices. On the other hand, some of the players doing well in Brazil point to the fact that by producing more components locally, the industry is less exposed to exchange-rate fluctuations.

Alstom's then Senior Vice-President for wind Alfonso Faubel told me in Rio de Janeiro that his company fully supports the new BNDES regulations, and that he is confident that the cost of producing turbines will come down in the medium term.

'The key thing is scale and an automotive-industry approach to production', said Faubel, who added that in terms of standard hours, the French-owned turbine manufacturer expected its Brazil nacelle assembly to reach very competitive levels in global terms in 2014.

Brazilian President Marcos Costa says 'the policy of [national development bank] BNDES is to localise production. In the short term, some components can become more expensive, but this is a learning curve, and we fully support the policy' (Backwell 2013c).

Political meltdown causes setback to wind industry

Brazil's annual wind installations reached a new record level in 2015 of 2.75GW, and were set to continue at around the same level in the following years, but the sector then found itself challenged by a political meltdown and the macro-economic slowdown that this caused. The meltdown began with widespread discontent around corruption and inequality around the 2014 World Cup and worsened as the 'Carwash' (Lavajatos) investigation into high-level kickbacks involving state oil company Petrobras and its big infrastructure projects widened and deepened. Carwash led to widespread arrests of politicians and company officials and eventually the impeachment of President Dilma Rousseff, who had been re-elected in October 2014. The ensuing crisis of confidence triggered an economic slowdown, a return to high levels of inflation and a crash in the value of the Brazilian Real.

At a conference in Rio de Janeiro in Autumn 2014, Mauricio Tolmasquim, then president of the Brazilian Energy Research Enterprise (EPE), and the architect of Brazil's successful long-term power auctions, described Brazil's booming wind-power sector as an "island of growth and optimism" amid a sea of political instability and a stalling economy. However, it was quickly shown that Brazil's renewable scene was not an island.

The wider political and economic crisis began to have an increasing effect on the economic variables that made Brazilian wind power a success, and the widening corruption investigation has dragged in major renewables investors. Financing became more problematic as the wider perception of risk grew.

The widening of the investigations to probe into construction contracts led to the powerful state development bank BNDES placing the parent companies of firms active in renewables on a blacklist due to the investigation into their parent company. These include Odebrecht Energia, and Queiroz Galvão Energia. Brazil's largest wind operator, CPFL Renováveis had construction company Camargo Correa – one of the companies at the centre of the Carwash allegations – as its largest private investor, although it has not yet been directly affected. As we have seen, the BNDES was providing the lion's share of financing for Brazil's wind industry and in some cases, an abrupt end to payments from the bank left companies with little alternative than to put their assets up for sale.

Brazil's economy stalled in 2014 after a decade of growth and then went into recession, due to combination of factors including: the end of a boom in commodity exports to China; the underlying low productivity and low investment rate of Brazilian industry; limits to spending by heavily indebted consumers; and the effects of Carwash. Exchange rate volatility and inflation – the traditional banes of the Brazilian economy returned to Brazil after more than a decade of relative stability and Brazil's credit rating was downgraded to junk status. Delays in payments from the BNDES and increased difficulties in obtaining bridging loans and finance for the remaining part of financing for projects began to lead to a squeeze on developers completing their projects, together with an erosion in wind-power projects' competitiveness. 'Getting financing during the current economic situation became very complicated', said GE's Head of Wind for Latin America Jean-Claude Fernand Robert. 'It is not happening in the time required to support power projects, and it is putting a tremendous stress into the system.' (Dezem 2016).

Economic slowdown eventually fed into government projection of energy demand and led to fewer auctions and a shrinking market, with power demand slumping by over four per cent in the first quarter of 2016 compared to a year before and industrial demand falling by 7.5 per cent. The culmination of this process came in 2016, when wind companies won no contracts in the April A-5 auction – with nearly all of the contracts awarded going to small hydro projects – due to uncertainties over the financial conditions they faced over the five year lead time, and the government cancelled the Reserve auction planned for December.

ABEEólica Executive Chairman Elbia Gannoum warned: 'There are big international companies that came to Brazil, set up factories and brought jobs. It is important to understand that, in the wind-energy market, energy contracted today will generate contracts and jobs for the factories in the next two years.' She added, 'Since we had a low contracting in 2015, we will have more idle factories by 2017. With zero contracting in 2016, we will have factories practically stationary in 2018. Big companies will not continue in Brazil with empty factories.' (Weston 2016a). A subsequent auction set for 2017 was also cancelled.

As we will see, US manufacturing giant GE acquired Alstom's energy business in 2015 and in Brazil this meant combining the number one and number two turbine suppliers in the Brazilian market, with the combined business accounting for about 40 per cent of supply. Feast turned to famine however in 2017, with GE Renewable Energy CEO Jerome Pecresse warning: 'This year we've had no volumes allocated. We have no visibility on future volumes to be contracted and there are very important delays in payments from customers (Dezem 2016).

GE said that Brazil needs about 1.5GW of contracting per year to support the existing turbine factories and related supply chain in Brazil. Bloomberg pointed out that Brazil had seven turbine factories in 2016, with each capable of producing around 400MW each – a total of 2.8GW. With orders slowing, GE reduced its workforce at its three manufacturing facilities in Brazil. 'There is a lot of pressure on renewables in Brazil right now', said Pecresse. 'That's scary.' (Dezem 2016).

It may be some time before demand for turbines in Brazil itself picks up, with the political crisis dragging on into 2017 and threatening to stymie economic recovery. In the meantime, companies manufacturing in Brazil may have to look to exports to other markets – growing markets in nearby countries like Chile, Argentina and Mexico in order to keep their factories running.

Meanwhile, the effect on the developer and wind-power operator market has been a wave of merger and acquisition activity, as local developers are forced to look for financing and foreign companies take advantage of a weak local currency and knock-down asset prices. As an example, dynamic Brazilian company Renova Energia found its ability to finance its projects challenged by the new situation and its debt growing, and so it sold part of its equity to ill-fated US-owned developer SunEdison, ahead of its collapse. At the time of writing it seemed likely that the company would end up being controlled by Canada's Brookfield, the company that eventually took over SunEdison and its TerraForm Global and TerraForm Power yield-cos (see Bautzer 2017).

Renova also sold its 386.1MW Alto Sertao II wind complex to US power company AES in order to be able to complete its 437MW Alto Sertao III wind complex.

Other highlights include the sale of Brazil's largest wind-power operator CPFL Renovaveis to China's state grid; the sale of projects developed by Casa dos Ventos to Canadian-owned clean energy and water infrastructure fund Cubico; a series of acquisitions by UK-based private equity Actis from Casa dos Ventos and Spain's Gestamp; and the sale of Queiroz Galvao Energia, which had built up one of the biggest portfolios of renewables assets before it was hit by the woes of its parent company, infrastructure giant Queiroz Galvao. At the time of writing, a number of other international companies were preparing to make bids for wind-power assets in Brazil, showing that despite the slowdown, investors continue to take a long view of Brazil's wind-power market.

The next wave

A number of other new markets have opened up to become sizable opportunities for the wind industry, including South Africa, Chile and Argentina. Let's look at these in turn.

South Africa

South Africa's electricity system is still dominated by coal, but a series of successful power tenders and rapidly declining costs has led to a strong policy momentum behind renewable energy and the opportunity for the country to establish itself as a regional leader.

From 2010, persistent electricity shortages contributed to South Africa's slowing economic growth and this allowed wind energy to emerge as an alternative to coal that was quick to install and easy to finance.

In 2011, the South African government set a renewable energy target of 18.8GW by 2030 including 9.2GW for wind, 8.4GW for solar photovoltaics (Solar PV) and 1.2GW for concentrated solar power (CSP), in its Integrated Resource Plan (IRP) 2010–2030. A draft IRP published in 2016 set the target for wind energy at 37.4GW by 2050.

From August 2011, South Africa has relied on a competitive bidding process, the Renewable Energy Independent Power Producer Procurement (REIPPP) programme to add renewables capacity. Under the REIPPP, projects are selected based on the bid-price (70%) as well as socio-economic factors (30%). Bid winners enter 20-year power purchase agreements (PPAs) with utility Eskom.

Four REIPPP Rounds have been held at the time of writing with a total of 6,391MW capacity awarded. Wind power has seen the biggest allocations so far, with a 53.3 per cent share, followed by Solar PV and CSP with 36.1 per cent and 9.4 per cent, respectively.

The REIPPP received international acclaim and attracted a host of international renewables players. The top three wind-turbine suppliers (Vestas, Siemens and Nordex) accounted for 65 per cent of capacity ordered or supplied for the first three rounds, while across the four REIPPP rounds, the leading wind developers were Enel Green Power, Mainstream Renewable Power/Actis and China Longyuan Power Group, which together accounted for 56 per cent and are all international players.

Between REIPPP Rounds 1 and 4, the average levelised power price fell 54.6 per cent for wind. According to the South African Wind Energy Association, wind power is now averaging R0.62 per kWh having fallen from R1.15 per kWh and R3.65 per kWh respectively in Round 1. This price is significantly lower than the tariff prices for coal from Independent Power Producers (R1.03 per kWh), Eskom coal (R1.05 per kWh to R1.16 per kWh) and nuclear power which is estimated at between R1.17 to R1.30 per kWh. (see SAWEA, www.sawea.org.za/index.php/media-room/press-releases/392-the-future-is-blowing-in-the-wind). This makes wind power the cheapest way to install new power generation capacity, even in a country with big indigenous coal reserves. Renewables deployment began to help alleviate the chronic power cuts that were affecting the country and prompted the South African government to expand allocations 100 per cent between Round 3 and Round 4. Thanks to the REIPPP programme, South Africa installed the largest amount of wind capacity in Africa in 2014 and ranked 14th globally in terms of annual installations.

So far so good. However in 2015 the country's fast-growing renewables sector began to face pushback from state-owned utility Eskom and elements of the ANC government. In principal, delays began to occur in Eskom in providing 'budget quotes' for grid connections – an essential element for companies that have been declared preferred bidders in bringing their projects to financial close. In some part this was due to the Eskom's own financial problems. Later, reports suggested that elements close to President Jacob Zuma were providing high-level cover to Eskom officials to delay the signing of PPAs for renewables projects due to commitments to a large-scale nuclear programme that had been proposed in conjunction with Russian nuclear firm Rosatom – with a reported cost of US$76bn. Rumours suggested that the turn towards prioritising nuclear was linked to the opportunities for 'state capture' of resources by some government officials.

Disagreement over the proposed nuclear programme was one of the factors behind the controversial firing of independent-minded Finance Minister Pravin Gordham in March 2017, and in April 2017 the Western Cape High Court ordered a halt to all procurement activities around the programme on the grounds that it was 'unconstitutional' (see Vecchiatto 2017).

The effective refusal of Eskom officials to sign PPAs led to delays in a number of projects awarded preferred bidder, and failure to bring the projects to financial close also led to knock-on delays in announcing the preferred bidders in the expedited, or 4X round, further stymieing much needed investment. Eskom officials seemed prepared to ignore directives from the government to get the agreements signed, and argued against the evidence that they were refusing to sign some of the contracts, saying that they were 'too expensive'.

Eskom's actions are 'in contravention of government policy', Mainstream Chief Executive Officer Eddie O'Connor said. 'These guys have gone completely rogue. It's increasing the risk of outside investors investing in South Africa fairly dramatically.' (Burkhardt and Cohen 2016).

Incumbent interests

Perhaps it is no surprise that wind power has received significant push back. South Africa ranks among the top10 global producers of coal, and coal continues to dominate South Africa's power generation. The REIPPP has benefited from delays in the construction of the 4.76GW Medupi and 4.8GW Kusile coal plants. Falling demand for coal threatens to cause coalmine closures and subsequent job losses, and so local demand from Eskom is seen as a way to sustain local coal industry jobs and the power miners' unions that support the ANC. South Africa's Department of Energy (SA DoE) has extended the IPP procurement programmes beyond renewable energy to include coal (2.5GW), among other sources.

South Africa is also among the world's top 15 producers of Uranium. The IRP 2010 includes an ambition to expand nuclear generation capacity by up to 9.6GW. The major obstacle is financing. However, offers by nuclear vendors such as Russia's Rosatom to provide project finance are providing the government with potential new options. In July 2015, the SA DoE and Rosatom signed two memoranda of understanding (MoUs): one to train personnel for the South African nuclear power industry and the other to enhance public awareness of nuclear energy in South Africa.

However, the arguments for both coal and nuclear have been critically under-mined by the falling price of renewable energy and technology development. The country's leading scientific research body, the CSIR (Council for Scientific and Industrial Research) published a landmark report in January 2016, which laid out a 'least cost' path for South Africa's energy development based on renewables plus gas fired generation and said that this option compared to nuclear and coal options would save the country some R100bn per year by 2040 (see CSIR 2017). 'Imagine what a future government could do with R100bn a year to invest in education, healthcare or job creation', said Mainstream CEO Eddie O'Connor. 'Decisions taken today could burden SA with crippling costs. That is why many companies are arguing for an electricity model that is fit for the future; one based on democratisation, decentralisation and division of Eskom's generation and grid businesses.' (O'Connor 2016a).

The scandal around Eskom's obstruction of renewables projects has done significant damage to the investment climate in South Africa. However, favour-able underlying economics and growing demands for more transparency in government mean that South Africa's path towards renewable energy will resume in the mid-term.

In February 2017, President Zuma publicly announced that the government was committed to the REIPPP and that Eskom would sign the PPAs. However, at the time of completing this new edition, the agreements had still not been signed.

Chile

Chile's small but dynamic economy is set to be increasingly powered by renewables in the coming years as government tenders open to all technologies have led to wind and solar projects winning market share from incumbent fossil fuel generation with extremely competitive price bids.

Chile's landmark August 2016 power tender resulted in renewable power generators ENEL/Endesa and Mainstream Renewable Power winning the lion's share of the 12.34 TW/h of supply on offer, at an average price of $47.59 per MW/h – a cost reduction of around 40 per cent compared to the previous tender.

Significantly, the tender pitted renewables projects directly against already built fossil fuel generation. Incumbents such as Engie Energia Chile, Colbun and AES Gener came away empty handed from the tender and saw their share prices tumble. All of these businesses and Endesa have large amounts of supply agreements due to expire between 2019 and 2022, but only Endesa is successfully replacing its fossil-fuel-based contracts with new renewable contracts (Endesa won 5.95 TWh of contracts in the August 2016 tender alone compared to the 6 TW/h of contracts due to expire).

Effectively, Chile's system is in transition from more expensive fossil generation to cheaper renewables. 'The result from Chile is the first example of a truly "open" auction, where renewables have competed directly with fossil generation – including fully amortised plant – and still won', said Mainstream Renewable Power CEO and Founder Eddie O'Connor (O'Connor 2016b).

Mainstream, which won around 30 per cent of the capacity on offer, is set to invest around $1.65bn to build seven wind farms with 985MW of capacity, according to Country Manager Bart Doyle. 'The auction puts a lot of pressure on conventional power and utilities now', said Doyle. 'It is a big wake up for the conventional utilities, to have a part of their portfolio in renewables' (Thomson *et al.* 2016).

One unique feature of Chile's system is that developers must offer to supply a certain amount of power at a given price, rather than offering capacity, and are obliged to go to the power spot market to purchase any shortfall if they are not able to do so. This poses new challenges for renewables generators, as they effect-ively need to supply 'firm power' from variable sources of generation, in the context of renewables becoming the dominant source of generation over the coming period.

Planning for 'renewable firm power' was an integral part of Mainstream's bidding strategy for the August 2016 tender. By generating precise wind measurement data for each project, Mainstream was able to build an accurate model of how much of the required energy each project would be able to deliver and how much energy each project would have to buy in the spot market. This in turn allowed it to bid at competitive prices in the tender, with the confidence that revenues would cover the cost of building the projects and the trading risk inherent in the contracts. The development of large-scale energy storage – whether using batteries, thermal energy, gravity or other technologies – will make

Figure 4.3 Mainstream Renewable Power CEO Eddie O'Connor meets Chilean President Michelle Bachelet, Santiago, 2016 (Source: Mainstream Renewable Power).

projects even more economically attractive in the future, Mainstream officials point out.

Also underpinning the tender prices are technological improvements at the production level. Better sensors and data-driven analytics, cheaper and lighter materials and bigger blades allow cutting edge renewable companies to further reduce the price of constructing new plants.

Another factor is scale. Renewable power supply works best when it benefits from geographic diversity. Each of the seven Mainstream wind projects has a different wind profile and therefore produces power at different times of the day. This will let the company combine output from the different projects to supply the energy required, reduce the time that no power is being generated by the projects, and thus reduce the need to trade in the spot market.

'Chile provides just a taste of what could be done in bigger and more diverse markets such as North America or Europe', says Mainstream's O'Connor. He adds:

> Increased penetration of renewables can create a virtuous circle that will continue to lower costs to the consumer as the cost of the technology falls further, and create more demand, while also bringing lower production and

financing costs thanks to increased scale. The idea of 'firm' renewables is no longer an oxymoron.

(O'Connor, personal communication)

Argentina

Argentina has long been seen as the 'wind-energy power that never was' due to it having some of the best wind resources in the world, unfulfilled power demand, a generation base made up of mainly ageing fossil fuel plants that has failed to keep up with demand, and the fact that it has been a net importer of gas for almost a decade.

However, attempts to kick-start large-scale wind development in the early part of the 2000s were largely thwarted by a government-imposed freeze on power prices that sparked a decade-long undeclared war with some of the country's major generators and distributors, which led to widespread underinvestment in the sector, while vested interests favoured incumbent power sources.

The state power company Enarsa built only a tiny fraction of the wind plants it planned in the early part of the last decade, and only a few projects – Genneia's 77.4MW wind farm in Rawson being the biggest – were successfully brought into operation.

The election of President Mauricio Macri in November 2015 ended more than a decade of rule by the Peronist couple of Nestor and Cristina Kirchner. Macri has declared Argentina 'open for business' again and moved towards restoring investor confidence in the country through rebuilding relationships with creditors, foreign governments and the local private sector.

Macri's assumption to power on 10 December 2015 opened up the path for renewables to take off in Argentina. His new energy team, led by former Shell Argentina President Juan Jose Aranguren made favourable statements on the need to favour wind power over other initiatives such as large-scale shale gas exploration and appointed former wind-power analyst Sebastian Kind as Sub-Secretary for Renewable Energies. Kind was an advisor for the Peronist Senator who introduced Law 27.191 (approved in September 2015), which foresees an increase in renewables to 20 per cent of power supply by 2025. President Macri confirmed the commitment of the new government to this target soon after taking power and began to take action to unwind Argentina's complex system of subsidies for power and other energy prices.

The urgency for Argentina in ensuring that its power system receives adequate investments could not be higher. On 22 January 2016, peak electricity demand reached a new high of 24.61GW during a heat wave that affected Buenos Aires, leading to a major power outage that affected around half a million customers. A few days later, the government announced that it would be raising power prices by up to 500 per cent for some user categories after more than a decade of prices having remained virtually unchanged.

Figure 4.4 GWEC delegation meets Argentinian Energy Minister Juan Jose Aranguren
and Renewable Energy Secretary Sebastian Kind in Buenos Aires, 2016
(Source: Ben Backwell).

In March, I helped the Global Wind Energy Council (GWEC) to organise a
visit to Argentina from a delegation of wind-power companies including Vestas,
Gamesa, Siemens, Mainstream Renewable Power, Acciona, Iberdrola, which
received red-carpet treatment from government officials at both a national and
regional level.

Kind and his team at the newly formed Renewable Energy Secretariat worked
around the clock through most of the year to create the framework for the first
of a series of planned tenders under the RenovAR programme, which were aimed
at creating around 10GW of new renewables capacity (or 20% of total
generation) by 2025, of which 60–70 per cent are expected to be wind.

The first tender and the additional 1.5 round contracted more than 1.4GW
of new wind-power capacity, with the lowest bid submitted for wind $46/

MWh in the first auction. The auction saw a mixture of local and international power producers and developers win acreage, including Argentinian companies with significant interests in existing gas-fired thermal generation such as Central Puerto and Pampa Energia along with local renewables player Genneia.

Chinese-owned wind-energy company Envision – which we have met in Chapter 3 of this book – was a surprise winner in the first auction, winning four of the projects on offer, adding up to 185MW out of a total of 708MW of wind projects awarded contracts. Envision Group Executive Director Felix Zhang said he expected Argentina 'to be one of the most promising emerging global wind markets'.

As well as continuing with a programme of regular auctions, Argentina also expects to see demand for renewables growing from private companies that can now contract directly with producers. One of the main features of Argentina's renewable energy law is a requirement that industrial consumers also obtain 8 per cent of their power from renewable sources in 2017 and 20 per cent by 2025, mirroring the national targets.

Realising Argentina's wind potential will not be all straightforward. The first two tenders in 2016 filled a large part of the country's spare transmission capacity and limited capacity transmission nodes in areas with good wind and solar sources could drive up prices in the next auctions.

The government was set to carry out a first tender for new transmission capacity in 2017 and it estimates that Argentina needed to add 5,000 kilometres (3,100 miles) of transmission lines in the next three years. More importantly, Argentina still needed to show that the chronic macroeconomic instability that has characterised most of its modern history is a thing of the past. The projects awarded contracts in the first two auction rounds were set to be mainly built using local or balance sheet finance.

However, building out future rounds will require companies operating in Argentina having access to competitive international financing in order to be able to bring prices down further and realise the potential of the country's wind sector.

5   The offshore frontier

By placing turbines in the sea, we can reap significantly higher levels of energy because of the strength and consistency of winds offshore. The resource is such that a relatively small area in the Northern European seas area could – theoretically – create enough energy to supply the entire world with electricity. The countries around the North Sea – Denmark, the UK, Germany and others – have been the first to take advantage of offshore wind's possibilities, but there are good resources offshore from the US to China, Japan, Korea and India.

Where offshore wind is developed is a function of economics, and countries need both good resources and a compelling reason to build wind power in the sea. Some, like the US, have good offshore wind resources, but there are abundant development opportunities for cheaper onshore wind. For others, such as the UK, offshore wind has become one of the only ways apparent to add zero-carbon power in large quantities – and relatively quickly – to replace its ageing generation base. The UK, which has become the undisputed leader in the sector, is fortunate to have probably one of the greatest resources of any single country in the world, with industry officials touting it as 'the Saudi Arabia of offshore wind'.

Harvesting this wind resource constitutes possibly one of the greatest engineering challenges that mankind has faced. Installing turbines in deep, rough waters means that they need to sit on stable foundations, and it is these that constitute one of the biggest costs and technical challenges. The other challenges are: installing the foundations and turbines in constantly changing weather conditions; ensuring that turbines actually work, and work reliably over long periods of time, withstanding constant strong winds, storms and corrosion; and connecting offshore wind farms to onshore electricity grids, requiring the building of offshore substations and the laying of large array cables on the seabed.

These challenges have inverted the economics of wind-farm development. In onshore projects, the wind turbine typically makes up 65–70 per cent of the cost of a project, while in a typical offshore project the turbine may constitute only 40–45 per cent of the cost, due to the high cost of the foundations and installation. The installation process and the cost of substructures are relatively inelastic at their lower range and so it makes sense to try to increase the amount of energy produced by each turbine. The result is a steady increase in turbine size, taking us to the 8MW Vestas turbine, which we mentioned at the beginning of the book, and beyond, with several companies now working on 10MW designs.

Beginnings

As we have seen, Henrik Stiesdal supervised the construction of the world's first offshore wind farm in 1990–91, 2.5km off the coast of Denmark at Vindeby. The wind farm was developed by Danish utility association, Elkraft, that later merged into DONG Energy, a company that would become one of the leading forces in the sector.

Until 2000, growth in offshore wind was slow, with development taking place on a small number of near-shore projects in Danish and Dutch waters, with turbines of less than 1MW capacity. The Middelgrunden project in Danish waters was the first large-scale project with twenty 2MW turbines, while seven 1.5MW turbines were connected to the grid off Utgrunden in Sweden the same year.

Since 2001, offshore wind has been steadily picking up momentum and the share of offshore in total annual wind-capacity installations has been growing, making it an attractive growth area for larger turbine manufacturers operating in a crowded onshore market. In 2001, the 50.5MW of capacity installed represented 1 per cent of total new European capacity. In 2012, offshore wind installed 1,166MW, representing 10 per cent of the European wind market total of 11.895GW. By June 2013, over 6GW of offshore wind had been installed.

The Vindeby project featured eleven 450kW turbines, giving it a total capacity of 4.95MW. Just over two decades later, in 2013, partners DONG, E.ON and Masdar were inaugurating the first phase of the London Array project, with 630MW of capacity, made up of 175 Siemens turbines each rated at 3.6MW. By 2016, the average turbine size had increased to 4.8MW and this will increase even more sharply in the coming few years as 7-10MW turbines become the "new normal".

While Denmark got things started in offshore wind, as it did in the modern wind market as a whole, and scaled up the size of its wind farms, the amount of potential it offers to developers and turbine manufacturers is dwarfed by its North Sea neighbour, the UK.

The prize: UK offshore wind

The UK is estimated to have over a third of Europe's offshore wind resource, and estimates give it a total theoretical potential in all waters of 120GW using only areas with water depths of less than 50m. The first offshore wind farm, Blyth Offshore, was built under the Non-Fossil Fuel Obligation (NFFO) and commissioned in 2000. It consisted of two 2MW Vestas turbines and was developed by a consortium including E.ON and Shell Renewables.

In 1998, the British Wind Energy Association (now RenewableUK) held discussions with the government and the Crown Estate, which owns almost all the UK coastline and its seabed. A set of guidelines were published that were intended to allow companies to develop wind farms of up to 10km² and 30 turbines, to allow developers to gain experience. In what became known as UK Round 1, 17 applications were given permission to proceed in April 2001.

The first Round 1 project was North Hoyle, completed in 2003, and since then ten more have been completed, with a total of 1.1GW. In December 2003, the Crown Estates announced the results of Round 2, with fifteen projects awarded and a total capacity of 7.2GW, with the biggest being the 1.2GW Triton Knoll area. Two Round 2 projects – Gunfleet Sands and Thanet – were completed in 2010, with several others following over the next couple of years, including London Array in early 2013. In May 2010, the Crown Estate gave approval for extensions to seven Round 1 and Round 2 sites, to allow the creation of another 2GW of capacity.

In the meantime, the Crown Estate had launched a third round of allocations in June 2008 that would dwarf anything that had come before, with potential for up to 33GW in nine zones. Bidding closed in March 2009 with over 40 applications, and with multiple applications for each zone on offer. The successful bidders were announced on 8 January 2010. The first planning permission applications for Round 3 zones were submitted in 2013, with a number of big power companies including Iberdrola, EDPR, SSE, Statoil, Statkraft RWE, E.ON, Mainstream/Siemens, Eneco and Centrica winning areas in different consortia, along with one small but highly aggressive specialist developer, SeaEnergy Renewables.

In parallel, the Scottish and the Crown Estate also called for bids for potential projects in Scottish territorial waters. Although these generally have deeper waters than in the other UK sites, seventeen companies submitted bids, and the Crown Estate signed exclusivity agreements for 6GW of sites with nine companies. Subsequently, five projects have been granted agreements for lease and several of these are now close to gaining planning consent.

The amount of potential GW on offer, and the Crown Estate's highly proactive role in trying to take forward development of the different zones, had large utilities and turbine manufacturers salivating. During the last two years of Gordon Brown's Labour Party government (2008–10), a number of turbine manufacturers including Vestas, Siemens, GE, Gamesa and Clipper had identified sites for manufacturing for new turbines that were meant to address the coming demand for UK projects.

Clipper began building a 4,000m² facility on Tyneside to manufacture the blades for its giant 10MW Britannia project in 2010. The same year, GE announced plans for a UK manufacturing facility with a planned £100m (US$169m) investment to create as many as 2,000 jobs. Siemens had announced plans for a £210m (US$354m) facility to build its new 6MW direct-drive turbine in Hull, while Vestas was planning a 70-hectare site in Sheerness in southeast England. Spain's Gamesa and France's Areva subsequently announced plans for manufacturing around the port of Leith in Scotland, while several other companies, including Alstom and REpower, were engaged in scoping sites for their facilities. A number of non-European companies, including Mitsubishi,

Figure 5.1 The giant London Array offshore wind farm in the Thames Estuary (Source: Mark Turner/London Array).

Samsung, Hyundai, Doosan and XEMC, also developed plans to deploy in the UK. Most of the talk was around a lack of available sites for testing new turbines and building manufacturing facilities, as well as actual and potential supply-chain bottlenecks, particularly in the areas of offshore cables and cable laying and in installation vessels.

The onset of harder times for utilities, followed by the new Conservative government's plans for wholesale reform of the electricity market (known as EMR), which involved scrapping the UK's highly successful Renewables Obligation and replacing it with a system based on contracts for difference (CfD), led to a significant slipping in the previously foreseen timelines for project development of Round 2/Round 1–2 extensions/Round 3 and Scottish Territorial areas. Greater uncertainty coincided with the onset of financial hard times for several turbine makers in particular, and most of the early plans for the creation of a UK offshore turbine industry were cancelled or delayed.

Clipper cancelled its plans for Britannia in August 2011 after UTC pulled the plug on the project. Although it never formally cancelled its plans in the UK, GE pulled back from any large-scale deployment of its planned 4.1MW direct-drive turbine after having second thoughts about both its platform and the market. Korea's Doosan fell by the wayside in early 2012.

Vestas was in crisis by the end of 2011, and, although it received planning permission for its Sheerness site in May 2012, it cancelled its plans a month later in June, leading to a messy dispute with the site owner, Peel Ports. The decision was linked to the squeeze on capital expenditure at the Danish company and the postponement of the deployment of what was then its planned 7MW turbine, as well as to its changing expectations over UK market growth.

Siemens, which had taken the lion's share of the Round 2 contracts and was quietly winning the contracts for most of the earliest-stage Round 3 projects, maintained its plans for its Green Port Hull facility, after having gained planning permission in May 2012. In March 2014, Siemens finally gave the go-ahead to start work on the factory after long negotiations with the UK government. Renewable UK chief executive Maria McCaffery said:

> It's the green-collar jobs game-changer that we've been waiting for. Attracting a major international company like Siemens to the UK, creating 1,000 jobs manufacturing turbines at two sites in Yorkshire, proves that we can bring the industrial benefits of offshore wind to Britain.
>
> (Lee 2014a)

Meanwhile, Spanish turbine manufacturer Gamesa had also been forced to slow down its offshore programme in 2012, relocating its 5MW prototype project from Virginia to the Canary Islands and putting its plans for a 7MW turbine to be engineered in the UK on the back burner. It did maintain interest in a site around Leith, however. Others, like Areva and Alstom, also slowed down their plans. The French turbine makers found that they had significantly more flexibility than some of their competitors, having won big in the French offshore tender in April 2012. The tender obliged both companies to create industrial facilities in France, and they will soon be in a position to supply some UK demand, particularly in the southern part of the country, from their French factories.

As of the end of 2013, UK offshore ambition in the next few years has been significantly scaled back, with the most optimistic forecasts giving a total of 13–16GW of capacity by 2020, compared to previous – rather over-ambitious even in the best-case scenario – estimates of up to 32GW.

As of 2016, the UK had total offshore wind capacity of 5.2GW and is set to reach around 11GW of capacity by 2020. While the UK did not see the wave of turbine manufacturers setting up factories that it expected at the start of the decade, both Siemens and MHI-Vestas have set up plants, in Hull and the Isle of Wight respectively, creating thousands of jobs. More importantly for the future, the industry has achieved a dramatic reduction in cost over the last decade, with the CfD auction held in 2016 set to see prices as low as £65–70/MWh, much lower than the Hinkley C nuclear plant and competitive with new build gas thermal plants. This puts offshore wind in a strong position to be the dominant source of large scale power supply in the UK in the decades to come.

Japan – the next big offshore market?

Outside Europe, one of the most promising markets for offshore wind is Japan, whose energy system is being transformed in the wake of the Fukushima Daiichi nuclear accident in March 2011. Analysts estimate that as much as 90GW of both fixed and floating turbines could be installed within the next ten years, and Japan's floating wind demonstration projects have given it a world-leading position in this emerging segment.

When Japan scrambled to shut down its 50 main nuclear power stations after the meltdown at Fukushima Daiichi in March 2011, it was left with a gaping 30 per cent shortfall in the production capacity needed to meet national electricity demand.

In the period between the accident and the end of 2013, the country has spent around 15tr yen (US$145bn) on imports of LNG (liquefied natural gas) to make up the output deficit, a running cost understatedly deemed 'unsustainable' by Shinz Abe's government.

There is still considerable uncertainty about how much of Japan's nuclear capacity will be brought back on line, with public opinion deeply suspicious about the technology. In a 2015 poll by the Japan Atomic Energy Relations Organization, 47.9 per cent of respondents said that nuclear energy should be abolished gradually and 14.8 per cent said that it should be abolished immediately. Only 10.1 per cent said that the use of nuclear energy should be maintained, and a mere 1.7 per cent said that it should be increased.

Another survey by the newspaper Asahi Shimbun in 2016 suggested that 57 per cent of the public opposed restarting existing nuclear power plants even if they satisfied new regulatory standards, and 73 per cent supported a phasing out of nuclear power, with 14 per cent advocating an immediate shutdown of all nuclear plants.

Despite this, the government of Shinzo Abe has backed a limited nuclear restart – reversing the position of the previous government and saying 'Japan cannot do without nuclear,' and set a target of obtaining between 20 per cent and 22 per cent of the country's power from nuclear sources by 2030 (see Suzuki 2017). As of early 2017 just three of Japan's 42 usable reactors were running, while Kansai utility Keptco had won permission to restart two more reactors after a legal battle with residents and NGOs around water contamination (see Hurst 2017).

Offshore wind is seen as the likeliest utility-scale energy resource to substantially reduce the country's reliance on nuclear. Japan has a rich wind resource off both its Pacific and continental coasts, estimated by the Ministry of Environment (MoE) to be equal to a theoretical capacity of 1,600GW. The island nation's narrow continental shelf means as much as 80 per cent of this potential lies over deep water, in depths greater than 50m, which is beyond the reach of conventional fixed-foundation technologies such as monopiles, jackets and concrete gravity-base solutions (CGBS).

The Japanese Wind Power Association estimates an upper-range figure of 519GW in floating offshore wind capacity, but a more realistic 141GW is thought to be economically harnessable in the medium term. In March 2014, the Japanese government introduced a new feed-in-tariff specifically for offshore wind projects, which was set at a generous Y36/KWh, and has subsequently maintained this level, while reducing tariffs for the burgeoning solar PV sector (see Foster 2016).

Four demonstration projects have so far been launched by the Japanese government through its R&D arm, the New Energy and Industrial Technology Development Organisation (Nedo), to test and fine-tune the leading technologies. At the heart of the initiative is Fukushima Forward. Backed by more than US$230m from the MoE, the 11-member consortium led by industrial giant Marubeni, and including Mitsubishi Heavy Industries, Japan Marine United and Shimizu, is studying the viability of a range of floating wind-power concepts through a cluster of four experimental units moored within sight of the stricken Daiichi nuclear power station.

During the summer of 2013, a 2MW Hitachi downwind wind turbine on a Mitsui semi-submersible foundation – dubbed Mirai – was floated out for grid connection, along with a 25MW 66kV floating substation, Kizuna, and installed in depths of 120m in the Pacific Ocean. In June 2015, Mitsubishi installed a 7MW MHT167/7.0 turbine (the turbine was renamed from its previous name of 'Sea Angel' so as 'not to confuse the market') on a three-column semi-submersible platform (see Smalley 2015). The pilot is the only multi-unit floating development off Japan and may ultimately lead to around 1GW of installations in the waters off Fukushima by 2018.

Off the western-most tip of Japan, near Nagasaki, a second pilot project was brought online in 2013 by a consortium including Fuji Heavy, Toda

Construction, Fuyo Ocean Development & Engineering and Kyoto University. The Goto demonstrator – a Hitachi 2MW on a novel 'super-hybrid' concrete spar hull moored in some 100m of water – replaced a scaled 100kW version of the turbine that had been floating on location since June 2012.

The extent to which offshore wind (along with other technologies like solar PV) fully or partially replaces Japan's nuclear capacity depends on the government's Basic Energy Plan, which is still in draft and is yet to be approved. Theoretically, replacing all the current nuclear capacity would require 90GW of offshore wind, with a further 60GW to take the place of the previously planned nuclear expansion. In other words, 150GW of installations could be the upper 'blue sky' range of offshore installation expectations over the next 16 years.

However, at least some of Japan's nuclear power stations are likely to be restarted – how many remains a moot question. It is also worth remembering that Japan is targeting more than 28GW of solar PV installations by 2020 and 53GW by 2030. Nevertheless, the offshore opportunity is likely to be considerable.

The Fukushima Forward and Goto pilots have been propelled into being in a very short span of time, and are emblematic of Japan's commitment to a potentially rapid expansion of floating wind power. Compare this to the rest of the world. The only installed utility-scale projects are Statoil's 2.3MW Hywind prototype off Norway (2009) and US start-up Principle Power's 2MW WindFloat off Portugal (2011). Both companies aim to build multi-turbine arrays using the same or similar technology, but these projects are still at the planning stage.

In floating offshore technology, the collaborative approach, with multi-industry R&D efforts and government-backed initiatives, should give Japanese developers, constructors, manufacturers and supply-chain participants a valuable industry-leading position. This could be exported to other potential markets, assuming that the technology can be proven and the cost of energy driven down. Opportunities in Japanese offshore wind may also open up for non-domestic companies, despite the market currently being something of a closed shop, with the major projects being built by consortia of the country's various industrial giants.

As we shall see, Mitsubishi Heavy Industries has merged its offshore business with Vestas, and the joint venture is likely to deploy the Danish turbine maker's 8MW V164 in Japanese waters, rather than the SeaAngel.

The Japanese floating wind projects are expensive, with a cost of energy of around 200–300 yen per kWh (about €1.40–2.10/US$1.95–2.90), compared to the most competitive European fixed-foundation installations at €0.11–0.12/kWh (US$0.15–0.16). However, it must be remembered that the pioneering Fukushima Forward and Goto projects were fast-tracked. Big savings are there to be had as the sector matures: it is an open secret in Japanese wind-power circles that the Fukushima consortium is looking to halve its cost of energy (CoE) for the project's second phase.

Much remains to be done in terms of regulation and in terms of engineering for Japan's offshore wind industry to really take off and development has been slower than had been hoped in the aftermath of Fukushima. However there is little doubt that offshore wind will eventually become an important part of Japan's energy mix.

Siemens cleans up

Before we look at which companies are in a position to win the leadership in offshore wind-turbine manufacturing, we need to look at how the contenders got to where they are at present.

According to Henrik Stiesdal, Siemens carried out the Bonus acquisition in 2004, as the Danish company was seen as one of the leading players in terms of quality and reliability, and because of its unique offshore experience. As well as building the first wind project at Vindeby, in 2003 Bonus had built what was then the world's largest project, the 166MW Nysted project with 72 turbines each rated at 2.3MW. In 2004, Siemens scaled up its turbine offering with a 3.6MW model – the SWT-3.6–107 – that was to prove the world's most popular offshore turbine and become for many financiers the only 'bankable' machine in the market, particularly between 2010 and 2012. A 4MW upgrade of the machine was unveiled in early 2013.

Siemens also invested in the installation process, buying 49 per cent of specialist vessel operator A2SEA in 2010 for DKK860m (US$160m). Siemens

Figure 5.2 Siemens' new direct-drive 6MW machine at the Gunfleet Sands offshore wind farm (Source: Siemens/Dong).

turbines have not been immune to quality glitches, such as a problem with corrosion protection that caused it to swap out the pitch bearings of its 3.6MW machines in 2010–11. However, the company proved that it had the operational strength to deal with the problem quickly and get turbines spinning again, and its reputation arguably emerged strengthened from the episode.

While the 3.6MW machine built a reputation for reliability, Stiesdal had become convinced that to make a permanent dent in both production and operations and maintenance costs, a new technology approach was required.

Siemens built its first direct-drive machine onshore in 2009, and began testing the prototype for its giant 6MW offshore turbine onshore in May 2011, with a second version with a larger 154m rotor diameter following in October 2012. In January 2013, Siemens installed two of the new 6MW turbines in an extension to DONG's Gunfleet Sands offshore wind farm in the UK. There are also plans to eventually build a 10MW machine.

Helped by the misfortunes that Vestas was suffering with its V90 machine, by 2009 Siemens had built its market share of offshore to 75 per cent, with partnerships with offshore-wind forerunners DONG Energy and E.ON playing a big role in its success. After a dip in 2010 when there was a surge in the number of Vestas turbines installed, Siemens' market share grew further and by 2013 it was 85 per cent. Even officials from Siemens say privately that this is 'not a healthy level' and they would like to see more competition. But that hasn't stopped the German company from hoovering up the lion's share of the most recently awarded offshore contracts.

Vestas gets burnt

One of the main reasons Siemens was able to become dominant in offshore is that the industry's biggest company, Vestas, has not been able so far to match its success onshore in the offshore space. Vestas constructed its first offshore wind farm, the 'Tunoe Knob' project in Denmark, as early as 1995, with ten V39 500kW turbines.

In 2001, it won the contract to supply 80 of its new V80 2MW turbines to construct Horns Rev, the first really big offshore wind farm in the North Sea. In December 2002, it finished installing the turbines ahead of schedule. However the turbines suffered a series of problems with transformers and generators over the following 18 months – badly insulated components were found to be the main problem – and all the 80 nacelles – plus the site's test nacelle – were taken back onshore for repairs in the summer of 2004.

Reports from the time describe a nightmarish experience. There was only one half-hour period during the whole 18-month period when all 80 machines were operating together. There were 75,000 maintenance trips – by helicopter – during the period, adding up to two per turbine per day. 'Experience is expensive, but also precious. Being the first large offshore project, Horns Rev must be a success', said Vestas' President and CEO Svend Sigaard. 'The project is important for Vestas' continued leadership in the offshore segment. It is my belief that Vestas will win the market in this segment. Even though it has been at a high premium, it puts Vestas and our suppliers into a unique position' (Vestas 2004).

The company had meanwhile won orders in the UK and the Netherlands. There is little doubt, however, that the issue caused lasting damage to the company's reputation in the offshore market, and worse was to come.

In early 2007, Vestas was forced to temporarily withdraw its V90-3.0MW offshore turbine from the market due to problems with the turbine's gearboxes, after 72 of a total of 96 V90-3.0MW turbines operating offshore developed major gearbox problems. It was only able to put it back on the market over a year later, on 1 May 2008. The V90 turbines that Vestas installed in Vattenfall's 90MW Kentish Flats project in 2005 were plagued with gearbox problems in their first four years of operations, until an upgraded version of the gearbox was installed in all the turbines in 2008 and 2009. The withdrawal from sale of the V90 delayed construction of Vattenfall's 300MW Thanet wind farm nearby and the £800–900m (US$1,350–1,500m) project, which became temporarily the world's largest when it was commissioned in 2010, also suffered initial gearbox problems after start-up.

Subsequently, however, V90s at Thanet and in the 180MW Robin Rigg wind farm in Scotland have performed well, according to the developers, with Robin Rigg developer E.ON reporting availability levels of over 98 per cent. The first units of Vestas' next offshore turbine, the V112-3.0, installed in E.ON's Kårehamn project in Sweden, have also been performing well.

However, although it was still picking up some orders for the V112, Vestas needed a big machine to get back in the game and compete for the big orders of the future. As we have seen, the company was forced to put back its initial plans to deploy and then mass-produce its 7MW V164 turbine in the UK. It then announced a slower-than-previous programme for building the V164 prototype and announced in 2012 that the turbine would now have an 8MW capacity instead of 7MW.

Building a prototype is one thing, but winning massive multi-billion-euro contracts from utilities for giant future wind farms in deeper-than-ever-before waters is another, and Vestas' executives had decided by early 2012 that they would need a heavyweight partner to really be able to compete.

REpower goes big

REpower, the German turbine manufacturer, was founded in 2001. In 2003, it launched the biggest offshore machine, the 5MW REpower 5M offshore wind turbine, at a time when the largest available model currently in the market was Siemens' 3.6MW machine. The company subsequently developed the 6M, with a 6.15MW rated power and deployed the first machine in 2009. As we saw in Chapter 4, the company was taken over in 2007 by India's Suzlon after a battle for control with French nuclear giant Areva, which has its own ambitions in offshore wind.

The first two 5M prototypes were deployed offshore in 2007 in the Beatrice demonstration wind farm in Scotland. Beatrice was a breakthrough project that deployed the turbines in deep water and supplied power to a nearby oilrig. It involved a team of people, including Ronnie Bonnar and Dan Flynn, that would subsequently go on to play important roles in developing big offshore projects in Scotland. REpower went on to build Vattenfall's Ormonde wind farm in the UK,

Figure 5.3 REpower's 5MW offshore turbine (Source: REpower).

the Thornton Bank project off Belgium and formed part of the Alpha Ventus demonstrator in Germany.

Former REpower CEO Fritz Vahrenholt went on to become the CEO of RWE Innogy. The high point of REpower's commercial success so far has been the signing of a memorandum of understanding in early 2009 with the German utility for the supply of 250 of its 5M and 6M offshore turbines up to 2016. The turbines are expected to be deployed in the Innogy Nordsee 1 and Nordsee Ost wind farms in Germany, as well as in the Netherlands and the UK.

In 2010, REpower hired Siemens Wind CEO Andreas Nauen away from its German rival; the company looked to be on a roll. By 2012, REpower had a cumulative share of the offshore installations of 8 per cent, in a market still dominated by Siemens and Vestas. However, it was responsible for 19 per cent of installations that year, its best showing yet.

However, since then orders have slowed, as the company suffered the effects of the financial problems afflicting parent company Suzlon. It lost an order for the Gode 1 wind farm to Siemens after developer PNE sold the project to DONG. And at the EWEA Offshore 2013 conference in Frankfurt, CEO Nauen warned that the company would have to start laying off workers if new orders did not materialise soon. This was only a day after unveiling its biggest-ever turbine, the REpower 6.2M152, and a few weeks after announcing that the company was changing its name to Senvion, ahead of the rights to the REpower name expiring.

REpower-Senvion had formidable strengths, having shown it could bring big multi-MW machines that perform reliably to market successfully, and ahead of its rivals. But there was a big question mark over its future as long as it belonged to Suzlon, and as long as Suzlon's finances remained under pressure. In 2015, Suzlon bowed to financial pressure and sold Senvion to US hedge fund Centrebridge, which carried out an IPO of Senvion shares in 2016. However, it remains to be seen at the time of writing if Centrebridge has the appetite or the patience to make the big investments that Senvion will need to make it to compete with its peers in the offshore space.

Enter the industrials

The long-term strategic potential of offshore wind attracted a number of heavy-weight contenders to the arena, whose major selling point is widespread industrial experience and knowhow and balance sheets big enough to offer developers security in case of major failure in any offshore wind project. Among European companies, two entrants were France's Areva and Alstom, who we met in Chapter 2.

Areva lost a battle to take over REpower in 2007 and instead bought Germany's Multibrid, which had its own 5MW turbine design. It successfully demonstrated its M500 5MW turbine in the Alpha Ventus project in Germany – dealing with a major third-party component failure in the process. With its HQ and production facilities in the German offshore-wind hotbed of Bremerhaven, it has since won a number of orders including the Global Tech 1 and Wikinger projects, and was also successful in the first French tender, winning the 500MW Saint-Brieuc zone. The company was also set to win its first UK order at the time of writing.

Areva hired highly respected industry veteran Julian Brown to run its UK operations in 2011, and began to look to building manufacturing facilities in Scotland, to add to its commitment to build a French turbine factory in Le Havre, and its existing Bremerhaven factory. Under pressure to compete with the big new machines planned by rivals, Areva announced in November 2013 that it would be bidding in the second French tender with a new 8MW design, and subsequently won 1GW of capacity with its consortium partners that included GDF Suez and EDPR.

Meanwhile Alstom deployed the offshore prototype of its 6MW Haliade machine in November off the coast of Belgium. Alstom won 1.5GW in the first French tender and committed itself to building two industrial hubs in France, while it has also identified manufacturing sites in the UK. Like Areva, having French manufacturing means that if it won projects in the UK or elsewhere in Europe it would be able to supply these from France, while waiting to see really big orders would be forthcoming from Round 3 projects. It hired former Vestas Offshore boss, Anders Søe-Jensen, a highly effective salesman, to work with its VP Frédéric Hendrick.

As well as Alstom and Areva, the South Korean chaebols also attempted to enter the offshore turbine market. Samsung installed the prototype of its S-7.0-171 7MW turbine – at that time the biggest in the world – at a test centre in Methil, northeast Scotland in the autumn of 2012, an extraordinary achievement for a company with a relatively short track record in the wind industry – and no experience at all in the offshore wind game. However, Samsung came very late to the offshore game, and its path to market was seen by most analysts as being by building up a project pipeline in its home market of South Korea and using its vast balance sheet to buy its way into projects in Europe and deploy its turbine in these.

In 2014, however, Samsung effectively cancelled its plans for offshore wind after months of rumours, and in 2015, its 7MW turbine prototype was sold to UK government technology accelerator ORE Catapult to be used as a part of a testing and training hub.

As well as Samsung, fellow chaebol Hyundai developed a 5.5MW turbine and was looking to enter European markets, although it also subsequently shelved plans for widespread deployment of the turbines. In addition, by 2014 there were a number of Chinese companies, led by Sinovel, with offshore turbines in the water and plans for giant turbines of 10MW and above.

But more significant than the Korean and Chinese players is the new joint venture that was announced in September 2013 by Mitsubishi Heavy Industries (MHI) and Vestas.

Vestas–Mitsubishi to challenge Siemens?

MHI had been developing its own offshore wind turbine, the 7MW Sea Angel using novel hydraulic technology, and had signed a framework agreement with UK utility SSE to both build and test a prototype and deploy the turbine in the developer's giant North Sea projects. This includes participation in the consortium that will develop the massive 9GW Dogger Bank Round 3 zone. But talks had begun with Vestas in early 2012 over a 'strategic cooperation' agreement. After a year of sometimes-tense negotiations, the deal was a reality. 'We have a clear strategic intent to become a global leader in the offshore wind industry', declared Vestas' buoyant Chief Executive, Anders Runevad, at the September announcement. The new company is in 'pole position' to win the race for market share, added MHI wind boss Jin Kato. The deal should propel both companies into the big time of the fast-growing offshore sector, creating a heavyweight competitor to undisputed market leader Siemens.

The agreement between MHI and Vestas revolved around the highly anticipated V164-8.0MW turbine (recently upgraded to a 9.5MW version). The development of the turbine was transferred to the new company – along with Vestas' order book for its V112-3.0MW offshore turbines; its existing offshore service contracts; and about 300 employees. In return, MHI injected €100m (US$135m) in cash and another €200m (US$275m) later, based on milestone achievements as the V164 was tested and brought to market. MHI will also bring considerable financial muscle and significant industrial expertise to the enterprise. 'Vestas needed a solid financial partner to compete with Siemens, Areva and Alstom on receiving future offshore turbine orders', says Patrik Setterberg, a senior analyst at Nordea, one of three financial companies advising Vestas on the deal. 'We argue that [the] Vestas/MHI JV will be a very strong alternative to Siemens if they successfully introduce the V164 turbine to market' (Backwell 2013e).

The talks, which were first confirmed by Vestas on 27 August 2012, took place – as we shall see in Chapter 7 – during the most difficult period in Vestas' history – a time when it faced a monumental financial hangover from rapid expansion, cost overruns, a cash-flow crunch and a major management upheaval. In late 2012, it seemed that creditors and banks were pushing the MHI deal as virtually the only way out for Vestas, while the disparity between the two companies' financial positions seemed to put the Danes at a huge disadvantage.

Industry sources say that the talks came close to being derailed over a number of issues – such as the value of Vestas' technology, who would control the JV and, according to some, MHI's attempts to gain access to Vestas' technology for its own onshore business, which has all but been destroyed by a series of patent disputes with GE. At times, there was intense scepticism over whether negotiations were even continuing. Officials from Vestas always maintained their hope that the MHI talks would succeed, while simultaneously trying to assure investors that there was a 'plan B'.

Vestas' Chairman Bert Nordberg – no stranger to Japanese companies after running Sony Mobile Communications – told me in March 2013: 'I was CEO of a Swedish–Japanese JV so I learned the same patience as the Japanese. This is a marriage, not an engagement or a love affair. I am stubborn and I don't want to sign something that isn't good for Vestas.' He added that the delayed V164 would be developed with or without the MHI deal. 'We are not so weak that we have to force it', he said.

MHI's enormous fiscal resources are a huge asset for the new JV. With annual revenues of 2.8trn yen (US$207.6bn) in its 2013 fiscal year, MHI will almost certainly have the financial muscle to be able to extend guarantees to large-scale projects using the JV's turbines, giving it a potential edge over many of its rivals.

MHI also brings a wealth of expertise – in areas such as industrial outsourcing and aerospace – not to mention the capacity to build and supply offshore wind installation vessels. The Japanese giant is also well placed to leverage the growing political and industrial momentum behind offshore wind in Japan, as part of the key Fukushima project consortium.

In terms of the immediate financial impact on Vestas, the deal is 'marginal', according to Vestas. But the announcement had an immediate effect in dispelling doubts about the sustainability of Vestas' existing offshore business. The new JV's chief executive, current Vestas Asia Pacific and China President Jens Tommerup, said at the Copenhagen press conference: 'The JV will give us a strong business case to get some projects for the V112 before the V164 is ready, so we will also be stronger in the short term.' Later that same day, Vestas received confirmation of a sale for 43 V112s to Dutch utility Eneco – a relief after a long period without a single offshore order.

The new joint venture should also be able to take advantage of existing relationships with major potential customers. On Vestas' side, Danish utility DONG, the world's largest offshore-wind operator, was involved in early testing of the V164, and became a key early customer for the joint venture. DONG has been Siemens' biggest offshore customer, and has been eagerly waiting for Vestas to provide competition. Vestas has also worked closely offshore with German utility E.ON and Swedish utility Vattenfall.

The venture got off to a good start. DONG announced MHI-Vestas Offshore Wind as its preferred supplier for the Burbo Bank Extension project in the UK, and the company subsequently won a series of other contracts in the UK, Germany and Denmark. MHI-Vestas successfully completed the commissioning of all 32 giant 8MW turbines in Burbo Bank in May 2017. In June the same year, it launched an upgraded 9.5MW version of the turbine, which was aimed at continuing to put pressure on its rival Siemens.

Consolidation takes off

The MHI-Vestas agreement was to be the catalyst for a wholesale consolidation of the offshore turbine manufacturing sector in Europe, which would leave just a handful of players intact to fight for a smaller than expected, but still substantial and growing offshore wind market. As analyst Robert Clover predicted:

> Aside from costs of R&D and the balance sheet capability to honour any warranty issues required, there are too few unallocated MWs in the Euro-

pean market for all the present aspirants to reach sufficient economic annual manufacturing scale to warrant the new investments required.

(Clover, personal communication)

In January 2014, Areva and Gamesa announced that they would be pooling their efforts in the offshore sphere and forming a new 50:50 joint venture called Adwen.

Both Gamesa and Areva had been looking at manufacturing sites in Scotland and, more importantly, both had a similar technology approach to their offshore turbines. Areva had sunk large amounts of cash into the offshore business with a so far meagre return, while, amid the wider crisis afflicting pure-play turbine companies in the 2011–2013 period (see Chapter 7) analysts long doubted Gamesa's ability to finance development of a large next-generation turbine, much less provide financial guarantees for multi-billion-euro offshore projects.

Some analysts questioned whether the benefits of a deal were even for the two parties. Areva already had a limited number of 5MW turbines in the water, was constructing two large German projects, had a contract for the 400MW Wikinger wind farm – also in Germany – and was a winner in the French offshore tender. It had a nacelle factory in Bremerhaven and a blade factory in Stade and was set to build another industrial complex in Le Havre.

In 2014, Gamesa had one 5MW prototype on a pier in the Canary Islands, and no orders. This does not tell the whole story, however. Areva's severe problems as a company in the stalling nuclear space meant that it's appetite for more capital expenditure in the offshore wind business had effectively ended, although this was not understood by the industry at the time.

Gamesa had a vast experience in onshore wind and runs one of the biggest turbine fleets in the world. Its engineering knowhow is world-class, as could be seen in its ground-breaking 4.5MW and 5MW onshore turbine platforms. It had considerable capacity in component supply – the JV will enter into a preferred supplier deal with Gamesa for some key parts. And it had a long-standing relationship with Iberdrola, the world's biggest wind-power developer, as its biggest shareholder.

Teams from Areva and Gamesa began work on comparing the two companies' respective 5MW turbines to identify the best aspects and components of both, while teams from Gamesa quickly began to contribute to Areva's planned 8MW machine, the development of which is already well advanced.

But this was not the end of the rush to consolidate. As we shall see in Chapter 8, in 2015 GE, which had abandoned its plans to industrialise a 4.1MW design it had acquired through taking over Scanwind, returned to the offshore wind sector through a massive takeover of Alstom Energy. The deal made the 6MW Alstom Haliade part of GE's portfolio, and it quickly moved to confirm Alstom's commitments to manufacture and supply turbines for the awarded French offshore tenders, and moved to install a 30MW wind farm in the US off Block Island, Massachusetts. GE seemed to signal that it was back in the offshore wind market for good, with officials hinting in 2016 that the company was working behind the scenes on a much bigger turbine, probably in the 10MW range. Meanwhile it moved to take closer control of the business division, replacing offshore wind-industry veteran Anders Søe-Jensen with John Lavelle in November 2016 (Søe-Jensen had replaced Frédéric Hendrick as VP for Offshore Wind in Alstom one year earlier).

In 2016 the high-profile and drawn-out merger between Siemens Wind and Gamesa came to a conclusion, creating a European wind-turbine mega-company that was set to occupy the number one spot in the world rankings. But the question of what to do with Adwen was a major complicating issue in the merger. Company officials worried that an acquisition could lead to competition concerns among EU authorities. Analysts pored over how Siemens would integrate Adwen, with its competing turbine portfolio and different approach to technology (Siemens has become increasingly wedded to a direct-drive approach, while Gamesa and Adwen use a mid-speed gearbox approach). Adwen was making a major investment in creating a giant, and highly anticipated 8MW turbine with a 180m rotor – the largest in the industry at the time of writing, and it was unclear where this would fit in with Siemens' offshore business which was busily creating its own 8MW turbine to compete with MHI-Vestas.

On the announcement of the Siemens-Gamesa merger in June 2016, Gamesa said that Areva had been given three months to decide whether it wanted to buy Gamesa's 50 per cent stake in Adwen or sell its 50 per cent share. There were attempts to sell the Adwen joint venture ahead of the Siemens-Gamesa merger. GE reportedly made an bid for some but not all of Adwen's assets. According to the reports GE was interested in some of Adwen's technology and its French offshore contracts, but not in Adwen's German operations and not in all of Adwen's commitments to build factories and create jobs in France, which would have left the US company facing significant duplication (see González *et al.* 2016). Eventually Gamesa announced that it was buying Areva's stake for a modest €60m in September 2016. The question of how Siemens-Gamesa will deal with Adwen's technology, projects and industrial commitments was still an open one at the time of the writing of the second edition of this book.

As of 2016 the market had consolidated into four Western turbine manufacturers: Siemens-Gamesa, MHI-Vestas, GE and Senvion, plus several Chinese players whose activity offshore was concentrated in their own market. This compares to over a dozen players that had announced major plans for the offshore sector just five years before.

The wave of consolidation started with companies failing – or deciding to abandon their plans for the sector – and ended with the mega mergers of 2015 and 2016. The consolidation was driven by the realisation that the addressable market for offshore wind turbines would be far smaller than anticipated in the heady days of 2008–09, when companies expected the UK alone to reach capacity of up to 50GW by the early 2020s, as well as balance sheet weakness and the need for cost cutting among some of the players in the market.

Consolidation has left the market with a small number of companies with formidable R&D and financial capabilities that were prepared for the next stage in the development of the offshore wind market – the driving down of costs to grid parity levels.

The price revolution

As we have seen, the first two decades of offshore wind saw an increase in turbine size to the 6MW–8MW range and the emergence of a small number of strong technologically sophisticated companies. The industry was still heavily dependent on government support, with prices of offshore wind in levelised cost of electricity (LCOE) terms significantly higher than prices in wholesale electricity markets or even prices for new-build gas or nuclear. This was to change dramatically – and sooner than analysts expected – in 2016 and 2017, as a whole new type of market emerged. 'Back in 2013 the industry was doing large projects and managing to finance these, but it was all very expensive', says Huub den Rooijen, Director of Energy, Minerals and Infrastructure at the Crown Estate (which manages the seabed around England, Wales and Northern Ireland). He explains: 'The nature of the game was saying to government "please support our industry so we can get costs down" and the industry was all about how to achieve cost reduction, collaboration, task forces'. 'The amazing thing that has happened is that these costs have come down and projects are being bid at costs that make it hard for conventional generation to compete', he goes on, adding 'we are now in a highly competitive world' (Den Rooijen, personal communication).

The catalyst for cost reduction has been a move to competitive auctioning in different European countries. These saw prices at €54.5/MWh in the Netherlands and €49.9/MWh Denmark. In Germany's 2017 offshore wind auction two companies DONG and EnBW made bids at effectively the same level as wholesale electricity prices, which are currently around €30/Mh.

'The zero subsidy bid is a breakthrough for the cost competitiveness of offshore wind and it demonstrates the technology's massive global growth potential as a cornerstone in the economically viable shift to green energy systems', said DONG's head of wind power, Samuel Leupold. However, he noted that a number of specific factors made it possible for the company to make its eye-catching bid, including the fact that companies do not have to pay grid connection costs (Clark 2017a). The schemes do not have to be fully up and running until 2024 and the bids show either an expectation that wholesale prices will rise, or that even more dramatic reductions in offshore wind's LCOE will be made by the time the projects start producing energy in 2025.

As Bloomberg reported, the subsidy-free bids were based on anticipation of the development of new machines by companies like Siemens that could have almost double the capacity of today's already huge turbines and keep taking advantage of the basic equation that bigger turbines can produce more energy while taking advantage of the same number of foundations and common grid connections, while reducing maintenance costs per MW. 'Dong and EnBW are banking on turbines that are three to four times bigger than those today', said Keegan Kruger, analyst at Bloomberg New Energy Finance. 'They will be crucial to bringing down the cost of energy.' (Shankleman *et al.* 2017).

In Denmark, authorities said that they are expanding offshore wind test sites to demonstrate turbines that will be as high as 330 metres, taller than the Eiffel Tower and the major turbine manufactures are all looking at machines that will have 10MW or more of capacity. 'The question of turbine capacity and wing span has never really been an issue from a technological perspective', said MHI Vestas joint CEO Jens Tommerup. 'We have already taken the capacity of our 8-megawatt platform to 9-megawatt. The real question is what can the market support.' (Shankleman *et al.* 2017).

Industry pioneer and former Siemens Windpower CTO Henrik Stiesdal points out that competition has forced offshore developers to revise their business models and become much more innovative: 'Until now the main modus operandi has been to base the next project on the previous one, and to shun any innovation other than increases in turbine size' (Stiesdal, personal communication). Project managers have been rewarded for sticking to the plans, no matter how costly, instead of being rewarded for being innovative. This has led to conservative methods, both technically and commercially, with acceptance of non-optimal technical solutions. In the new business environment all suggestions for cost savings are taken seriously, and project managers are now rewarded for being innovative.

Stiesdal is sanguine about worries that price reductions are going too fast for the offshore wind industry, and that this could lead to project failure and/or manufacturing becoming unprofitable. 'I don't think there is a danger that this goes too far too fast', he says. 'It is the opposite that is the case. Because of the dramatic changes in cost, offshore volumes will go up, and this will make the low-cost bets safer.' Stiesdal predicts that offshore wind turbines will go to 10MW and above. However he challenges the idea that the market trajectory towards bigger and bigger turbines will always be the best way forward. 'It may well be that mid-sized turbines, which are generally more competitive on the turbine side due to larger volumes and a higher degree of industrialization, will re-enter the market once the infrastructure costs decrease sufficiently', he says, predicting that the biggest steps offshore will come on the infrastructure side, with industrialised foundations and rationalised electrical infrastructures (Stiesdal, personal communication).

The impact of the price revolution

The arrival of cost-competitive offshore wind will have a profound effect on European and global energy markets.

In Germany, the arrival of large-scale, zero-emissions, subsidy-free power from offshore wind will allow it to put its Energiewende on a sustainable footing and eliminate coal power generation.

In the UK, the expected allocation of contracts at prices significantly below the price awarded to EDF's Hinkley C nuclear project (£92.50/MWh) is set to have a major impact on debate around the future of the energy market and on the government's industrial strategy.

Huub den Rooijen says: 'The interesting thing about offshore wind for governments is that it lends itself to indigenisation, so you get local industrial activity, the production of components, O&M services. So you get a societal element as well as cheap low carbon energy.'

The path to truly significant integration of offshore wind requires new thinking about grids and markets, but this is by no means a bad thing, as offshore wind will help to accelerate development of a twenty-first-century power system. Offshore wind developers, equipment providers and stakeholders such as regulators are having to take a 'whole system' approach and look at a grid that will integrate a very large amount of variable generation, buffered by some gas generation and leaning increasingly on technologies such as storage solutions, the potential that electric vehicles bring and demand-side response technologies and business models.

Just as is happening with solar PV and onshore wind, offshore wind generators will find themselves dealing directly with customers as they look to find off-takers for their power among the growing market of companies who are looking to hedge their power supplies both from a price point of view and from the point of view of carbon emissions and sustainability impacts. 'So there are a billion new opportunities, as offshore wind drives innovation right into the sweet spot for UK and Europe', says Den Rooijen, adding 'suddenly we are talking about a technology that looks close to perfect.'

What will drive this process will be regulatory reform, as governments respond to the new technology possibilities and try and unleash all the innovation that is occurring. Of course, huge challenges still lie ahead. The new, low-cost projects still need to be constructed and operated, and there will still be issues around the environment and the cumulative impact of the new giant wind farms to manage.

In addition, the question of politics will continue to be centre stage. The UK has been the undisputed leader of the sector and has the strongest resource of any country. But its regulatory set up has some clear disadvantages compared to Germany, the most prominent being that the German government effectively plans and pays for grid connections for offshore, while the UK has opted for the rather complex system of discrete, privately financed and managed connections known as OFTOs.

The UK is also facing a period of self-inflicted uncertainty around its links with Europe because of the BREXIT vote and the protracted negotiations around exiting the EU. The UK had been a strong advocate of the creation of a common EU power market, and was increasingly moving to a system model based on increased interconnection capacity to allow more power trade. Offshore wind in the North Sea – an area that has the energy potential to supply the world's entire power demand many times over sorely needs a coordinated international approach to be developed in an optimised fashion through the building of inter-connectors and initiatives like the North Sea Wind Power Hub in Dogger Bank proposed by Dutch and Danish grid operators TenneT and Energinet.

However, it is likely to surpass the UK as the global leader in terms of additional offshore wind in 2017–2026, although the UK will retain the number one position in cumulative offshore wind installations at the end of that period.

Future growth in the offshore market

While 2016 marked a slowdown in global offshore wind installations, the market was expected to rebound strongly in 2017 and is likely to grow from 4,032MW in 2017 to 7,240MW in 2026, a compound annual growth of 13.8 per cent across 10 years, according to FTI Consulting.

Europe will continue to take the lead in 2017–2026, and is likely to account for 59.8 per cent of the new capacity added worldwide, followed by Asia Pacific (30.7%), and North America (9.5%). For this to happen, non-European wind markets will account for a large part of the growth over the 2017–2026 period.

New markets

Asia

China installed a modest 592MW of offshore wind capacity in 2016. However, it is likely to surpass the UK as the global leader in terms of total added offshore

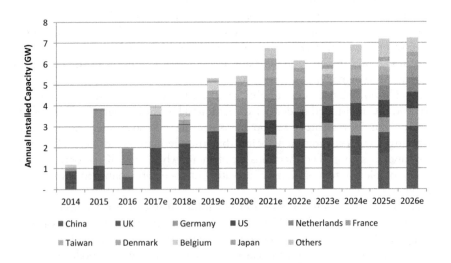

Figure 5.4 Near-term global offshore demand is downgraded, but medium-term looks promising (Source FTI Intelligence).

wind in 2017–2026, although the UK will retain the number one position in cumulative offshore wind installation by the end of 2026. China's offshore wind has gone slower than analysts expected and the Chinese NEA reduced its 2020 grid-connected offshore target to 5GW, while aiming to have a further 10GW under construction by then. Government support remains strong, and to ensure the stable development of the local offshore wind market, the NDRC has not only kept interim FiTs unchanged for both nearshore and inter-tidal projects (despite that the onshore wind FiT has been reduced twice in the past three years).

Meanwhile, Japan launched its first offshore wind-power auction in 2016 after the government amended its existing port and harbour law to promote offshore wind development. The Japanese Ministry of Economy, Trade and Industry (METI) has retained its offshore wind feed-in tariff of 36/kWh (€0.28/kWh), the highest available anywhere for offshore wind. However, local developer Marubeni has cancelled projects and reduced project sizes due to the challenge of low profitability. Meanwhile, Japan's New Energy and Industrial Technology Development Organization (NEDO) continues to support the development of floating offshore wind turbines with an aim to lower the cost of energy by JPY 20/kWh (approx. EUR 0.16/kWh) after 2030.

Taiwan has a very ambitious offshore wind target, 4GW by 2030, one-third higher than the original plan set in 2013 by the Ministry of Economic Affairs (MOEA). At the time of writing the second edition of this book, only two turbines, totalling 8MW, had been installed, but the industry consortium led by local shipbuilder CSBC has launched a strategic alliance aimed at helping Taiwan reach the target of 520MW by 2020 and 3GW by 2025. Foreign investors are keen on investing in Taiwanese offshore wind and in January 2017, DONG Energy and Macquarie Capital acquired a combined 85 per cent stake in the Formosa 1 project from Swancor Renewables.

The United States

In North America, most wind industry participants consider that the offshore wind sector is finally about to take off, although progress is still being made in stops and starts. The US connected its first offshore wind project – the 30MW Block Island using 6MW GE turbines – off the coast of Massachusetts in 2016, but 2016 also saw the US Department of Energy pull the plug on further investments into Principle Power's 30MW WindFloat Pacific project off Oregon, Dominion Virginia Power's 12MW VOWTAP project off Virginia and Fishermen's 24MW Atlantic City Wind Farm project off the coast of New Jersey due to those projects not meeting the established development milestones.

Only two demonstration offshore wind projects, totalling 33MW, are expected to be built before 2020, but the market is ready to take off from 2020 onwards. As of early 2017, more than 20 offshore wind projects, about 16GW in total, were under various stages of development in the US and there is strong support for offshore wind among key politicians from states on the Eastern Seaboard.

Massachusetts signed new energy legislation that calls for the development of 1,600MW of offshore wind capacity by 2027 into law in August 2016.

Offshore wind is part of New York Governor Andrew Cuomo's strategy, which aims at 50 per cent of New York's electricity coming from renewable energy sources by 2030 and 80 per cent by 2050. In his State of the State address in January 2017, he committed to New York adding 2,400MW of offshore wind capacity by 2030.

There is a growing appetite to pick up acreage for projects in the US from international developers. Apart from leading European offshore wind developer DONG Energy – who acquired two offshore wind leases with a potential for more than 2GW offshore wind capacity – in the US, Norway's Statoil won its first offshore wind lease that could accommodate more than 1GW of offshore wind in December 2016 and other major European developer utilities such as Iberdrola and Innogy were also entering the US market. It has taken longer than expected for the US offshore wind market to take off, but the future looks bright.

6 After Copenhagen

A perfect storm for turbine manufacturers

Standing for several hours in sub-zero temperatures in the snow surrounded by riot police can be a sobering experience at the best of times. Journalists are no strangers to chaotic scenes – or queuing for that matter – but when I realised that members of the negotiating delegations attending the December 2009 Copenhagen climate talks were standing behind me in the same crowd I started to realise something had gone quite wrong with the organisation of the talks.

Even more sobering were the subsequent events inside the conference centre, when it became clear that the Danish hosts had underestimated not just the logistics of managing what had become a political and media circus, but the underlying difficulties of reaching any meaningful accord on a binding climate deal. The efforts of Denmark's Climate Minister Connie Hedegaard, German Chancellor Angela Merkel, the UK Energy and Climate Change Secretary Ed Miliband, Brazil's 'Lula' da Silva and France's Nicolas Sarkozy came up against the hard fact that neither China nor the US – the world's number one and number two carbon emitters, respectively – were interested in a deal. Christian Kjaer says:

> Leading up to Copenhagen, we were all quite optimistic. It was during Secretary of State Hillary Clinton's speech in the first week of Copenhagen that made us realise that the Obama administration was not on board and that the Danish Presidency of the COP had not been able to reconcile the positions.
>
> (Kjaer, personal communication)

A whole range of attitudes just added to the noise: the emerging countries – encouraged by China – putting their efforts into an extension of the inadequate Kyoto agreement; farcical interventions by supposed climate 'radical' (and major oil producer) Venezuela and its allies; NGOs missing the point of what was happening; and the heavy-handed tactics of Denmark's small but pit-bull-like security apparatus.

It did not help the process when the competent and experienced Danish Climate and Energy Minister Connie Hedegaard handed over the chairmanship to her newly appointed Prime Minister Lars Løkke Rasmussen for the second week of negotiations. Neither did it help that the Swedish EU Presidency negotiated a position that had not been agreed by the other EU member states. But COP 15 would have failed, even in the absence of these challenges. China and the United States did not come to Copenhagen with the intention of achieving a climate deal for the period after Kyoto and one has to wonder why Chinese Premier Wen Jiabao and US President Barack Obama showed up at all.

Even at the event itself, some wind-power companies were providing bullish forecasts of growth in the event of an agreement, something that was widely expected by the greater part of the industry. 'With a very strong support of the Copenhagen meeting, and a very clear road map from political leaders around the world, I think I can deliver by 2020 a $50bn size of the business', said Suzlon's Chairman Tanti (Stromsta 2009). Suzlon notched up sales of 260.8bn rupees (US$5.6bn) in the fiscal year ending 31 March 2009.

Tanti estimated that global turbine sales, which had averaged 28 per cent annual growth over the decade prior to Copenhagen, would jump to 35 per cent a year in the event of an agreement, a prediction that was far higher than the 22 per cent growth expected by GWEC.

Others were more wary. Faced with increasing signs that little progress had been made in negotiations ahead of the summit, Vestas' CEO Ditlev Engel said 'Let's see what happens in the coming days, I am still hopeful we can get things going' (Backwell 2009a). Engel highlighted the forthcoming presentation of EU countries' national action plans, and China's massive wind plans, as important factors that would keep the industry growing fast, even if an agreement at Copenhagen was not reached.

A few weeks later, industry leaders were putting a brave face on the failure of the talks, but it wouldn't be long before the wind sector was finding that wind directions had changed. After years of fast expansion, and a sellers' market where turbine companies could not ship enough turbines to keep customers supplied, the tables had turned.

Well before Copenhagen, the effects of the 'sub-prime' financial crisis that hit the world headlines with the collapse of Lehman Brothers in September 2008, were making themselves felt. Bank lending for wind-power projects had become extremely restricted, while companies found refinancing their debts increasingly difficult.

Bloomberg New Energy Finance estimated that by early 2009, investment in renewable energy was down 50 per cent from its peak a year earlier. 'The amount of capital available to finance projects shrivelled to nearly nothing, as liquidity problems made banks either stop lending for infrastructure altogether or demand tougher terms, including shorter payback conditions', said REN21's Virginia Sonntag-O'Brien (GWEC 2009: 5).

Any money still left became very expensive, with financiers becoming more risk averse than usual … the amount of equity a project developer had to provide to secure a loan increased dramatically in comparison with the pre-crisis days, when projects could be financed with as much as 90% debt.

(GWEC 2009: 5)

Making things worse, stock prices had dropped dramatically, preventing companies from raising money through new equity issues, while the recession, which resulted from the financial crisis, was slowing down power demand. Meanwhile, natural-gas prices were falling.

Significantly, the financial crisis was also hitting wind projects hard in the US, with the disappearance from the market of most 'tax equity' investors – typically big banks or corporations – who invest in projects to profit from federal tax credits. The financial crisis meant less profit and, hence, less need for tax credits. This meant that wind developers could effectively no longer make use of the Production Tax Credit (PTC), the main vehicle for US government support of the wind industry, even though it had been extended until the end of 2012.

Financing rebounded by the end of the year in great measure due to government stimulus programmes – in the US, for example, the government made it possible to convert the PTC tax credit into up-front investment grants – and multilaterals. Investment in clean energy was around US$145bn, only 6.5 per cent lower than the previous year's record figure (subsequent revisions suggest that in fact the dip in investment may have been exaggerated at the time).

And the wind industry continued to grow fast in 2009, by around 41.5 per cent compared to installations the year before and 32 per cent on a cumulative basis, or 38GW of new capacity, mainly due to one factor: China.

China's wind industry added a staggering – at least back then – 13.8GW of wind, more than doubling its capacity to 25.8GW, and overtaking Germany to become the world's second-largest wind-power market in cumulative terms behind the US. Bloomberg New Energy Finance reported 'extraordinary investment activity' in China in the last quarter of 2009. China accounted for 72 per cent of the US$4.7 billion raised through initial public offerings (IPOs) in the sector, and for one-third of the entire increase in globally added wind capacity.

For the international turbine manufacturers, however, China-centric growth was little relief, given that, as we have seen, they were effectively shut out of most of the growth and that prices in the Chinese market were at extremely low levels. In what looked like an ominous development for the mainly European incumbents, 2009 was the year that two Chinese companies, Sinovel and Goldwind, muscled their way into the top five global wind-turbine suppliers. And while companies like Vestas were able to post record financial results in 2009, orders for turbines – which are typically delivered 6–18 months later – were falling.

Europe slows while China growth skyrockets

In Europe, sub-prime had largely morphed into the Eurozone crisis in early 2010, with Greece making its first request for a bailout on 23 April, and Spain and Portugal suffering crises over the following months (Ireland had already begun the latest phase of its long-running financial crisis in late 2009). Other Eurozone economies were feeling differing degrees of pressure from net lenders such as Germany, to large but more bankable debtors such as the UK and Italy. Soon, governments in some of the fastest-growing wind markets were slamming the brakes on after years of spectacular growth, while others were thinking about how to reduce costs, even if this meant slowing down the transition to clean energy.

In 2010, the annual wind market decreased slightly for the first time in two decades, with the industry adding 38.3GW, or 0.5 per cent less than the previous year. Once again, China installed a new annual record of capacity – 18.9GW – and overtook the US as the world's biggest wind power in cumulative terms. Also for the first time, more new wind power was installed in developing and emerging economies than in traditional OECD wind markets.

In contrast to the continued ramp-up in Asia, the US wind market was only half the size of the previous year, at 5.1GW.

Europe's wind market was down by 10 per cent in annual terms at 9.91GW, with EU countries accounting for 9.29GW of the total. While the still small offshore market grew by 51 per cent, the onshore market decreased by 13 per cent. Eastern European countries were responsible for a large proportion of new installations, with Romania installing 448MW and Poland 382MW. Among the established wind powers, Spain was the largest market, adding 1.5GW, with Germany just behind, and France, the UK and Italy following. In January 2011, EWEA CEO Christian Kjaer commented:

> Remarkable growth in the onshore wind markets of Romania, Poland and Bulgaria could not make up for the decline in new onshore installations in Spain, Germany and the UK during 2010. These figures are a warning that we cannot take for granted the continued financing of renewable energy.
>
> (EWEA 2010)

In 2011, the pattern continued. The market rebounded somewhat, with 40.5GW of installations, 6 per cent higher than the year before. Once again, the majority of installations were outside the OECD and GWEC said the trend 'now seems firmly established' (GWEC 2011).

Global clean-energy investment reached a new record of US$260bn, with the key driver being a big rise in public-sector investment as part of government stimulus packages. State-owned development banks and agencies were among the top asset finance arrangers for clean-energy projects, including the US Federal Financing Bank (US$10.14bn), Brazil's BNDES (US$4.23bn), German development bank KfW, the Nordic Investment Bank, Danish EKF, the European Investment Bank, the World Bank and many others (see table in GWEC 2011: 25).

China once again saw massive growth – 17.63GW – but this was less than in 2010. Analysts noted that the double- or triple-digit growth rates in capacity that the market had seen for nearly a decade were gone, and that the market was entering a phase of consolidation, with substantial excess manufacturing capacity. The decisive factor behind slowing growth in China was a lack of grid infrastructure to address the massive growth in the wind sector over the previous few years (see Chapter 3), and this would be a major factor holding back global rates of growth in 2011–13.

The US rebounded slightly, adding 6.81GW or 30 per cent more than installations in 2010. Momentum was once again building, in spite of – or because of – the uncertainty over whether PTC support would continue beyond the end of 2012. Significantly, the US industry was able to supply about 60 per cent of the manufacturing content for wind installations, compared to 25 per cent a few years before. This was a testament in part to the number of European companies that had set up new plants in the US in the last couple of years, including Vestas, Gamesa, Nordex, Acciona, REpower and Siemens. Capturing part of North American demand was key to the globalising strategies of these companies, but was clearly fuelling excess capacity throughout the industry.

Europe's installations were almost unchanged at 10.28GW, as the Eurozone crisis continued to stifle growth. Germany, the continent's most steady market, led with 2.08GW, followed by the UK (1.29GW of which 752MW was offshore), followed by the 2009 leader Spain, then Italy, France, Sweden and Romania.

A number of EU governments introduced measures to reduce support for renewable energy or began to look for measures to do so. As GWEC noted:

> On the one hand the EU has its 20-20-20 targets and on the other hand a budgetary crisis ... A troika comprised of the International Monetary Fund, the European Central Bank and the European Commission is watching over the budgets of Greece, Ireland, Portugal and Spain, and have in writing advised the Portuguese government to stop funding renewables via its budget ... This highlights the political risk involved in a political market.
>
> (GWEC 2011)

Spanish market's demise shakes the market

Spain's financial crisis and economic slump led to a rapid deterioration in political support for renewables, as the number one priority became reducing the fiscal deficit. While the solar PV and then wind sectors, as the fastest-growing energy sources, bore the brunt of political attacks, it is worth pointing out that Spain's energy demand had stopped growing and gone into decline, meaning that the country did not need new generation capacity of any sort.

Over the course of 2011 and 2012, the government introduced a series of measures, including a 'pre-register' that restricted the number of eligible projects that could still be built before the current regulatory scheme expired at the end of 2012. This was followed by a one-year moratorium on new projects – which in practice was extended indefinitely due to a lack of a new regulatory system for 2013 onwards, and a number of negotiated and unilaterally imposed changes to the feed-in tariff system. This meant that while there was still growth of more than 1GW in 2011 and 2012, these were effectively the last projects, and manufacturers were no longer clocking up sales from mid-2011 onwards.

Finally, in early 2013, the Spanish government imposed a new – highly unfavourable – regulatory scheme in the face of outright opposition from the industry.

The abrupt turnaround in the Spanish wind industry's fortunes had a major impact on the European industry. Spain had installed as much as 3.55GW in 2007 and 2.46GW in 2009. By 2011, the annual level had slumped to just over 1GW and by 2013 installations had practically dried up altogether.

The Spanish wind sector also had reverberations throughout the global industry. First, massive spare capacity was created from 2010 onwards, when the turbines for the last batch of projects were produced. Spanish manufacturers Gamesa and Acciona suddenly found they had factories without markets, as did Vestas, Alstom and Enercon. Gamesa, in particular, had to practically reinvent itself to survive, as we shall see in a separate section.

Second, the giant Spanish utilities that had pioneered the global expansion of the wind industry, Iberdrola, Acciona and – Portuguese but Madrid-based – EDPR, found themselves fighting a rearguard action against retroactive changes that would impact their earnings from their already constructed projects, while also seeing their capacity to raise finance affected because of Spain's situation. In the case of Iberdrola, the company found itself having to face hard choices in its political lobbying as the government laid out its reform programme. Iberdrola was one of the two biggest Spanish power companies owed government money as part of the €25bn (US$35bn) so-called 'tariff deficit', a huge sum of money accumulated on utilities' balance sheets as a result of a long-running mismatch between accepted generation costs and artificially low consumer tariffs. With its interests in fossil fuel, nuclear and large hydro-generation, Iberdrola began adopting a high-profile stance as a critic of Spain's solar industry, and began to qualify its support for continuing expansion of wind power in Spain. From being recognised as the champion of Spanish wind energy, the company found itself in violent disagreement with the rest of the Spanish wind-energy sector over its role in the recent legislative changes.

With its balance sheet severely constrained, Iberdrola also had to make hard decisions about how it was to maintain its position as the world's biggest wind-power developer. The company had already begun to slow its rate of growth in 2011, mainly because of decreasing possibilities in signing power purchase agreements (PPAs) for new US projects.

The company had then attempted to shift growth into emerging European markets, and in particular Romania. But growth in Europe was harder to create than it had foreseen, because of delays in getting big projects off the ground and the growing climate of austerity. And finally, in 2012, the company began a programme of selling 'non-core' assets in some onshore wind markets in order to make an ambitious gamble on harnessing growth from offshore wind. In the future, the UK (including offshore) was to be the primary destination for Iberdrola's wind development, along with some growth markets such as Latin America.

Overall, the Spanish crisis had a profound impact on the competitive landscape of the wind industry, as manufacturers lived or died on their ability to win market share outside Spain, along with an exodus of skilled wind executives seeking jobs elsewhere. The reasons why Spain's government stopped wind-power growth are understandable, but short-sighted. Power demand is no longer growing and the government had been in a situation of practical insolvency. However, the country continued to subsidise its coal industry as well as importing expensive fossil fuels, making a continued steady development of wind a good idea in the medium term. Managing a relative slowdown of the sector could have been handled differently in a way that recognised the strategic leadership position that Spanish companies had established – rivalling their Danish counterparts – in many areas.

Instead, the Spanish government has effectively taken a wrecking ball to one of its only high-tech growth industries, causing untold damage in terms of flight of expertise and capital. The fall-out had an impact far beyond the borders of Spain: 'If it can happen in Spain, it can happen in any country is the logical conclusion made by wind-energy financiers. They react by increasing the risk premium on financing everywhere', says Kjaer (personal communication).

US hits new peak, China slows

With European growth having levelled off, market growth depended more than ever on the US and China, together with some relatively small, but fast-growing emerging markets.

Continued uncertainty over the PTC and US government support for emissions reductions generally had the perverse effect of producing the biggest year ever for US wind installations in 2012, as developers rushed to install their projects before the threatened expiry of the scheme.

Installations were 13.2GW, with the US once again becoming the world's biggest annual market. Wind was the biggest source of new power capacity in the US – ahead of nuclear and thermal power – for the first time ever. As we have noted, the significance of this cannot be overestimated, given that gas prices were hovering at around US$3–4/MBTU throughout the period. Bloomberg New Energy Finance also noted that the majority of wind projects were built in states without renewable portfolio standards, meaning that utilities were buying wind power because they wanted to, not because they had to. On the other hand, massive growth in 2012 – with 8.4GW in the last quarter alone as the PTC deadline neared – was bound to set the scene for a big 'hangover' year in 2013, in a typical manifestation of the boom–bust cycle that short-term PTC programmes have created for more than a decade. US volatility has been one of the most difficult issues for wind-turbine manufacturers to handle, as illustrated by Vestas, which was laying off workers and considering plant closures in the US in a year of record wind-farm installations.

On the other side of the Pacific in China, the grid curtailment issue and the effects of too-fast expansion on inventories and company balances hit home hard in 2012, largely cancelling out the surge in US growth. China added 'only' 12.96GW, ceding the top place in annual installations to the US.

European growth picked up slightly, with 12.74GW of installations. Wind energy represented 26 per cent of all new EU power capacity, with wind meeting 7 per cent of Europe's power demand by the end of the year. However, as EWEA CEO, Christian Kjaer pointed out, the figures reflect 'orders made before the wave of political uncertainty that has swept across Europe since 2011. I expect this instability to be far more apparent in 2013 and 2014 installation levels' (EWEA 2013b).

In total, the market grew by 44.8GW, 10 per cent higher than the level of installations in 2011.

Wind shows its resilience

The wind industry faced significant setbacks in 2009–13, after almost a decade of outstripping the most optimistic projections. Overall, however, what is remarkable is not that the rate of growth of annual wind installations slowed during the period, or that there was a high level of volatility, but that wind power has been so resilient.

Underneath the yearly ups and downs, what is notable is that wind was competing, even when the most decisive factors – levels of government support and the comparative price of fossil fuels – were moving unfavourably. The real success story for wind, and the one that will ensure its long-term survival, is that it has become competitive with fossil fuels. Reaching the holy grail of grid parity is not clear-cut, due to the jumbled field of government support for renewables and fossil fuels, but it is clear that wind was close to reaching some kind of competitive parity in many markets around the world. As we have seen, in some markets, including Brazil, India, South Africa and some parts of the US, wind was already directly competitive and as we shall see in Chapter 9, this would soon be the case in a growing number of markets.

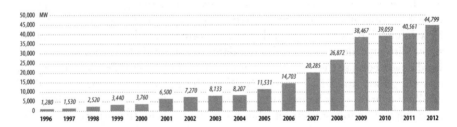

Figure 6.1 Global annual wind-power installations, 1996–2012 (Source: GWEC).

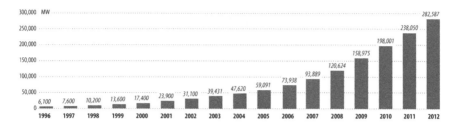

Figure 6.2 Global cumulative wind-power installations, 1996–2012 (Source: GWEC).

Ironically, the wind-energy industry in 2009–13 was experiencing its hardest challenge at a time when the technology was the most cost-competitive it has ever been. For the first time, wind industry officials were able to argue that the biggest challenge is no longer one of competitiveness – at least when it comes to onshore wind – the challenge is removing subsidies for all energy sources and creating a level playing field.

As the then IEA Chief Economist Fatih Birol (now Executive Director) called fossil-fuel – not renewables – subsidies "Public Enemy No1" at a speech given to the wind industry in Vienna in early 2013. He noted that the current global total in fossil-fuel subsidies for 2011, was $523 billion, an incentive equivalent to $110 per ton of carbon emitted. In comparison, global subsidies for renewables amounted to $88 billion in the same year.

Birol was right, of course, but as I predicted in the first edition of this book, only a further sustained and generalised fall in the cost of wind power would be capable of pushing wind to a tipping point where sustained rapid growth levels are assured.

In the meantime, however, turbine manufactures that had spent heavily to meet the next expected boom in sales were entering very troubled waters.

7 Turbine manufacturers in trouble

Vestas hits bottom

I flew over to Copenhagen on a bitterly cold day – 12 January 2012 – after being summoned along with other journalists for a major announcement on the restructuring of Vestas. Some changes had already been announced – or leaked – in the previous days, and it was clear that there was going to be blood on the carpet.

The backdrop to the restructuring was a series of worsening results, combined with two consecutive 'profit warnings' – the last in January – revealing big under-estimates of losses, which wiped out the company's 2011 profits and shocked investors. 'You have a company with little cost control, inadequate capital control and in some of their factories no operational control. It is business school 101 on how not to run a company', was the way one analyst, Martin Prozesky at Bernstein Research, described Vestas in 2012 (cited in Milne 2012).

After almost a decade of constant – even reckless – expansion, the seriousness of shrinking returns and slower than expected wind-market growth had finally hit home.

CEO Ditlev Engel announced in a sombre tone that 2,335 jobs would be cut, with the possibility of a further 1,600 being lost in the US. Whole units of the company were shut down and folded into each other, as the company took an axe to its costs and planned investments. The old geographical-based sales units were placed under a central global sales division.

Technology R&D was broken up, with parts of it going to the new Global Solutions and Services division – tasked with increasing revenues from outside the core area of turbine sales – and a new Turbines division headed by former Technology R&D Vice-President Anders Vedel. Global Solutions and Services also absorbed the existing Spare Parts and Repair unit. The Towers, Control Systems, Nacelles and Components and Blades units were merged into the new Manufacturing division. Wind, where Vestas had been pouring resources into its planned 7MW turbine and new production facilities, was also closed down as a separate division. The changes represented a significant streamlining of the management structure, with the number of executive managers being reduced from fourteen to six.

The casualties among Vestas' senior management, meanwhile, included offshore head, Anders Søe-Jensen; technology research and development (R&D) head, Finn Strøm Madsen; control systems head, Bjarne Ravn Sørensen; and head of communications, Peter Kruse.

Beyond the announcements, speculation centred around the fact that the man responsible – at least in formal terms – for getting Vestas' projected results so badly wrong and issuing the profit warnings – CFO Henrik Nørremark – was being promoted to an even more powerful position in the company.

The restructuring envisioned a new six-person management committee, with Nørremark as Chief Operating Officer and deputy CEO. The CFO position remained unfulfilled while the management board carried out an executive search for the right person. CEO Engel, whose leadership had been the subject of growing criticism, denied rumours that he had offered to resign at any point during the events of the previous months, saying 'The time to leave is not when there are challenges but when things are going well' (Backwell 2012a).

Most of the blame for the company's trouble seemed to be apportioned to the company's R&D department, which had underestimated costs on the development of Vestas' flagship V112 wind turbine and its Gridsteamer technology. Engel claimed that blame could be attached to Nørremark for the delays in receiving revenues in the last quarter of 2011 and for a series of cost overruns, saying 'All the challenges that have surrounded Vestas in the last weeks basically come from one area: the implementation of new technology' (Backwell 2012a). When asked about the logic of promoting Nørremark into such a powerful position in the company, Engel said he could 'not think of a better person to be COO' (ibid.).

The elevation of Henrik Nørremark proved to be short-lived, and reportedly represented a desperate compromise by two men who were engaged in a long-running power struggle. In the struggle, Engel, a relative newcomer to Vestas, had the support of key institutional investors, while Nørremark was one of a group of company originals from Jutland, which also included Central Europe President Hans Jörn Rieks and Jan Pilgaard.

Insiders say that the board was extremely sceptical about the elevation of Nørremark – and Chairman Bend Carlsen had made disparaging remarks about the company's 'accountants' prior to the restructuring – and was wondering how they could sell such a move to the company's investors, who had just collectively lost their shirts following the profit warnings. Market reaction, a key auditor's report, and the realisation that more bad news was on its way pushed the board and Engel to make a new move.

On 7 February, Nørremark was unceremoniously dismissed. A statement from the company said: 'The board of directors of Vestas Wind Systems A/S has … received a thorough briefing on the conditions which during the last months have led to profit warnings … As a consequence of this, CFO and deputy CEO, Henrik Nørremark, resigns' (Backwell 2012b).

Ditlev Engel took over his responsibilities directly, joking that he was now 'CEO, CFO and CTO' (conference call with analysts and journalists on 8

February 2012), while the search for new executives continued. No explanations were given for Nørremark's dismissal at the time, and further bloodletting was taking place. Søe-Jensen's replacement, Rieks, had been dismissed at the same time as Nørremark – once Engel had decided to move against Nørremark it was time to move against the whole group. His successor was former Ricardo and Areva executive Karl John, although his tenure was to prove short-lived. And on a wider level Vestas' programme of job cutting continued with white-collar and factory redundancies, aimed at reaching a level of 19,000 employees at the end of the year, from its 23,000-plus peak in 2010.

The backdrop to the management struggle going on at Vestas was a steady fall in the company's share price – Vestas shares fell 90 per cent between peaking in 2008 at 692 kroner (US$129) and the 58 kroner (US$11) level they had reached at the time of the restructuring. They were to fall further to a low of 23.25 kroner (US$4.3) on 20 November. In late 2012, up to 30 per cent of the company's shares were held by so-called 'short-sellers', 'investors' who had bought shares in the expectation that the price would fall further.

Along with successive waves of job losses in Denmark that the company announced in 2010 and 2011, Ditlev Engel had gone from poster child to villain in a short amount of time.

Relations with Vestas' syndicate of key bank lenders had also deteriorated, with the company announcing that it had not been able to fulfil its half-yearly financial covenant agreements when it reported its second-quarter financial results in July. Rumours began to circulate that the banks could force Vestas into a new share issue, something that could have been highly damaging to existing shareholders – and risky – in the prevailing negative climate.

Behind the scenes, Vestas' board was negotiating with Swedish telecoms executive Bert Nordberg to take over from long-running Chairman Carlsen, and it nominated him on 16 February. Nordberg became chief executive of Sony Ericsson in 2009 and steered the sale of Eriksson's 50 per cent stake in the Sony Ericsson mobile telephone joint venture to Sony in early 2012, becoming first CEO and then Chairman of Sony Mobile Communications. He was recommended by already-serving board member Håkan Ericsson, who had worked with Nordberg at Ericsson in Silicon Valley. He carried out a wholesale realignment of Sony Ericsson's product offering, and steered the venture through its takeover by Sony. But the company was also unable to regain significant market share under his leadership.

The appointment of someone with extensive experience of doing business in Japan and with the Japanese was almost certainly no accident. Nordberg took over the reins at the company's AGM on 29 March. On 24 August, Nordberg said that the company would seek a long-term strategic investor that could take a stake of up to 20 per cent in the company. On 28 August, Vestas announced that it was in talks with Japanese conglomerate MHI over a possible 'strategic cooperation' agreement. Although talks were known to centre on the capital-expenditure-demanding offshore wind sector, analysts speculated that the talks could lead to MHI taking a significant role in Vestas, or even taking over the

Figure 7.1 Vestas' Chairman, Bert Nordberg (Source: Vestas).

company entirely, despite Nordberg saying, 'Vestas is Danish, and shall remain a Danish company listed on the Danish stock exchange. That is my ambition and that is the company's strategy. It is important for Vestas and also for Denmark' (Lee and Jensen 2012).

The news was received extremely positively by analysts and Vestas' lenders, who imagined a possible cash injection into Vestas that could head off the threat of any new share issue. Many of them probably didn't imagine how long negoti-ations with MHI would go on, and the highly secret talks became the focus for jittery investor sentiment over much of the next year. Several times, negotiations came close to the brink over a number of issues – such as the value of Vestas' technology, who would control the JV and, according to some, MHI's attempts to gain access to Vestas' technology for its own onshore business, which had all but been destroyed by a series of patent disputes with GE.

Nordberg told me in March 2012:

> I was CEO of a Swedish–Japanese JV so I learned the same patience as the Japanese. This is a marriage, not an engagement or a love affair. I am stubborn and I don't want to sign something that isn't good for Vestas.
>
> (Nordberg, personal communication)

He added that the delayed V164 would be developed with or without the MHI deal. 'We are not so weak that we have to force it', he said.

The talks were a cause of considerable anxiety, however. In February 2013 in Tokyo, Mitsubishi Wind boss Jin Kato seemed to deliberately ruffle feathers by telling me in Tokyo that the company was set to deliver 700 of its 7MW Sea Angel turbines – a clear rival to the V164 – to Scottish utility SSE for deployment in UK Round 3 projects. Kato and Vestas' then boss Ditlev Engel seemed at pains to avoid contact during the same Tokyo event.

Aside from overseeing negotiations with MHI, Nordberg's main moves were continuing with job losses, trying to recruit high-level individuals to Vestas' depleted executive management and taking a broom to the company's internal problems. By making sure that the rest of the bad news lurking in the closet came out during Ditlev Engel's tenure – and I would argue that his reputation had suffered almost irreparable damage by this time – and that the most important job cuts were made quickly, Nordberg was ensuring that his successor would get a clear run and be able to bask in the glory of a good-news flow once the worst of the company's problems were behind it. Nordberg was also under pressure to make sure that any hint of financial impropriety had been dealt with in order to convince the intense due diligence that any deal with MHI would imply.

Nordberg appointed a fellow Swede Dag Andresen, a former Vattenfall executive, as CFO in April and Jean Marc Lechene, a veteran of the tyre manufacturer Michelin, as Chief Operating Officer at the end of June. The investor day held by Vestas in Aarhus on 3 October was all about 'cost out' – the process of running a rule over all of the company's operations in order to get production back to its original cost estimate – and other measures being taken to regain a positive margin.

Vestas attempted to balance the primarily conservative message of the investor day with the announcement that its planned 7MW V164 would now be an even larger 8MW. But there was no escaping the fact that there had been a clear reframing of the timescale for offshore within Vestas, and that it was doing everything possible to minimise and spread out capital investment on the new machine.

Vestas was able to meet its bank covenants and agree a new set of financial facilities in late November. It was clear that stability was returning as lenders could see its cost savings and cash-management programme advance, even though third-quarter numbers had been hair-raising, with free cash flow coming in at a negative €142m (negative US$181m), compared with a positive €276m (US$379m) in the same quarter of 2011.

On 7 November 2013, the company announced an 'intensification' of its cost-cutting programme that would see it shed an extra 2,000 jobs in 2013 and save an additional €150m (US$193m), bringing annual cost-savings to €400m (US$555m) since the end of 2011. The latest round of job cuts left it with a global staff of 16,000 by the end of 2013.

Chasing the dragon

Vestas' restructuring programme had clearly put the company on the track towards breaking even in 2013 and a return to profitability in 2014.

Meanwhile, the talks with MHI continued, despite frequent rumours that the talks had ended without success, or that Vestas' lenders were trying to force a deal with the Japanese by the beginning of 2013.

However, the management battle within Vestas had still not played itself out, nor was the position of Ditlev Engel, the 'last man standing' of the company, by any means secure.

On 2 October 2012, Vestas announced that it had terminated Henrik Nørremark's severance package in September after learning that he had entered into 'two commercial agreements' in India on behalf of the company without informing other senior management, or the board, causing the company a loss of '€4m (US$5.17m), possibly up to €18m' (Backwell 2012c). Nørremark contested the claims and began legal proceedings over the issue. At the time, Vestas did not go into details over what the Indian deals involved, and Nordberg said that a 'complex investigation' was under way (ibid.).

The following May I reported that the allegations involved a mysterious attempt to buy wind turbines in India from Chinese competitors. Nørremark is alleged to have set up a top-secret project called 'Project Dragon', by which Vestas' R&D department in India was to have set up a 100MW wind farm involving an unspecified number of Chinese machines.

Officials from Vestas told me that Vestas' Contract Review Committee, which approves all turbine contracts over 10MW, was not informed of the deal, and nor was Vestas' CEO Ditlev Engel. Apparently, Vestas did not take delivery of any turbines, nor was any wind farm constructed. Nordberg said in a statement that the inquiry by the company's auditors and lawyers 'proved that neither the board nor the group president and CEO have been involved in or have had any knowledge of the mentioned transactions' (Backwell 2013g).

In July, Vestas filed two lawsuits against its former Indian joint venture partner RRB at the High Court of New Delhi, after starting arbitration proceedings in the International Chamber of Commerce in June, in an attempt to recover €24m (US$32m).

Indian company RRB Energy and its London-born owner Rakesh Bakshi were Vestas' original joint-venture partners in India. The two companies parted ways in 2006 when Vestas sold its 49 per cent stake in Vestas RRB after acquiring NEG Micon, which had its own presence in India. However, they maintained close ties, with Vestas agreeing to focus on 750kW-plus turbines and RRB concentrating on smaller machines. Vestas agreed to continue to assist RRB through a technological cooperation agreement. Danish papers highlighted a private dinner that took place between Vestas' CEO and Bakshi in the summer of 2008, although this was three years before the disputed payments were made.

The true nature of Project Dragon has remained a mystery.

In February 2016, Vestas issued a statement saying that

> After successful negotiations among the parties, a confidential settlement has been concluded, which is conclusive and resolves all issues in dispute in multiple jurisdictions. Consequently, the trial in Denmark and the pending

cases in India, including the arbitration proceedings, have been and will be withdrawn.

(Vestas statement, 25 February 2016)

Vestas Chairman Nordberg stated: 'A settlement has now been reached between the parties regarding the mentioned disputes which is agreeable to all parties'.

The worst is over?

By mid-2013, Vestas seemed to have weathered the storm. The company had announced orders of over 2GW as of early July, with the second-quarter intake the strongest quarterly figure since 2011. Patrik Setterberg, senior analyst at Nordea, said that the company's strengthening order intake was 'not a coincidence, but rather a sign that Vestas' customers have greater trust in the financial position of the company' (Backwell 2013h).

Analysts MAKE pointed out that Vestas was in the best position to take advantage of a small global increase in orders – there was an increase of 48 per cent in volume compared to the same period in 2012. 'On the data we have seen for the first half, Vestas is performing the most consistently across the three regions', said Research Director Robert Clover (Backwell 2013i).

Investment bank HSBC noted that Vestas, along with other turbine manufacturers, had taken 'bold steps' to restructure its business and that the cost of restructuring measures had already been booked in 2012.

Meanwhile, Vestas' share price had recovered by 270 per cent in July from its low point in summer 2012, helped by investment banks like HSBC and Nordea putting 'buy' or 'overweight' recommendations on the stock. By early December 2013, the shares were back up to around 150 kroner (US$28), almost 400 per cent higher than a year earlier.

'A future without surprises'

The recovery in Vestas' financial fortunes was not enough to save Ditlev Engel, however. On 21 August the company announced that Engel was leaving and would be replaced by Anders Runevad, another Swede, and another former executive from Ericsson and Sony Ericsson.

While thanking Engel for his work over the previous eight years, Chairman Nordberg said, 'We feel that after eight and a half years we are looking for a tweak of the culture and more focus on profitability, and a future without surprises' (Milne 2013). The remarks were polite but cutting, given that in his last few years as CEO Engel had presided over a management war, four profit warnings and at least two big investor lawsuits against the company.

Vestas' shares rose 5 per cent on the news of the new CEO. The message from Runevad was clear: management would be focused on the sober tasks of efficiency and cost management, in close liaison with the company's major shareholders.

The move was a classic in ruthless management. Vestas' board had effectively waited until all the bad news was out of the way, and the bulk of unpopular job cuts had been made, before 'terminating the terminator', in the words of one former Vestas official who had lost his job back in 2012. The *Financial Times* cited a person 'close to' Bert Nordberg as saying 'For us it was important that we weren't in the most vulnerable situation. He was looking for a time when the situation was more stable so you could give more time to the new chief executive to settle in' (Milne 2013).

The more generous analysts pointed out that Engel had been better at building up Vestas than in managing it when times got tougher. Others criticised Engel for pushing ahead with big global expansion plans at a time when the global economy was about to go into recession and renewable-energy support schemes were threatened, while its investments in China soon became a liability as the country's own manufacturers took over the market. 'A different management skill set is required to run a business for growth than to restructure a business, and the two skill sets are often not present in the same individuals', says Robert Clover (personal communication) who followed Engel's fortunes closely as a leading industry equity analyst. It was also reported that institutional investors that had lost big on Vestas' shares during the crisis period would not come back while Engel was in charge.

Runevad is set now to preside over a period of positive developments. But whatever Engel got wrong, the wind industry lost one its most charismatic figures when he resigned, and one who had been determined to propel wind into the big leagues of international energy business and politics.

Mitsubishi comes through

A few weeks later, on 27 September, Vestas and MHI announced the creation of an offshore joint venture at closely coordinated early-morning press conferences in Copenhagen and Tokyo. It was over a year after talks were first announced. Officials from the two companies said that they expected the new joint venture to challenge Siemens for the top spot in the offshore wind business (see Chapter 5).

Back to profit

When Vestas released its full 2013 results in early February 2014, it seemed to have delivered on the main areas of its two-year turnaround plan. The company pulled in full-year revenue of €6.084bn (US$8.44bn) in 2013 and its earnings before interest and tax (EBIT), before special items, was €211m (US$293m), with a free cash flow of €1.009bn (US$1.4bn). The EBIT figures were 44 per cent better than analysts' consensus expectations.

Vestas said that 'the higher-than-expected revenue and EBIT were primarily driven by a smooth execution in terms of installation and transfer of risk combined with favourable weather conditions in December' (Lee 2014b). For 2014, it expected revenue to amount to a minimum of €6bn (US$8.3bn) with an EBIT margin before special items of at least 5 per cent and a minimum free cash flow of €300m (US$416m).

The figures showed how deep Vestas' cost-cutting measures had bitten. Annualised fixed capacity costs had been lowered by €484m (US$671m) compared to the fourth quarter of 2011. Net investments had shrunk by more than €500m (US$690m) to €239m (US331$m) since 2011 and working capital had been lowered to negative €596m (US$826m).

Vestas ended 2013 with orders of 5.96GW, up from 2012's 3.74GW. It delivered 4.86GW against 6.04GW. 'A double-digit EBIT margin in the fourth quarter and a free cash-flow generation of more than €1bn in 2013 are major achievements for Vestas', said CEO Anders Runevad (Lee 2014b).

What next for Vestas?

The restructuring and the exit of most of the Vestas 'historicals', the MHI deal and Engel's departure marked the end an era for the company.

The emphasis was now on efficiency and financial management. Structural problems remained; increased competition form big industrial companies like GE and Siemens; a slower market in Europe, difficulties in competing in China, and a cyclical US market that could leave it with big overcapacity. However, the company continued to have formidable strengths; a geographical diversity second to none that allowed it to benefit from the appearance of new markets around the world; a huge resource of knowledge and experience of wind energy; strong products and the advantage that having the largest existing portfolio of turbines brings; and the backing of government in the world's most wind friendly country.

As we shall see, the advantages ensured that Vestas was able to bounce back with a vengeance as strong market growth returned.

Rivals feel the pain

Of course, Vestas was not the only company that suffered during 2011–12. Hard times hit companies in different ways and, as we have noted, there was some considerable shelter for companies that were part of larger industrial groups.

Life was undoubtedly the most difficult for publicly quoted wind-power specialists. One – Clipper, the second-largest turbine producer in the US – teetered on the edge of bankruptcy, was bought out by a larger group and then shut down.

Gamesa adapts to Spanish meltdown

Another company, Gamesa, had to make an even tougher transition than Vestas. Gamesa had been one of the biggest beneficiaries of both the Spanish wind boom and – linked to this – the expansion of the world's largest wind developer, Iberdrola. As we have seen, Iberdrola is also the company's biggest shareholder with a stake that has oscillated around 15–20 per cent of the total. The two companies had successive framework agreements that meant that up to 50 per cent of Gamesa's business was coming from orders from Iberdrola.

Iberdrola's annual installations reached a high point of 1.78GW in 2010, but installations slowed down after that. The Spanish energy crisis stopped development there and it slowed down its breakneck expansion in the US in 2011 as it found it harder to sign new PPAs. Attempts to make up the shortfall by speeding up in Europe – notably in Romania and elsewhere in Eastern Europe – came up against transmission and regulatory obstacles, while its balance sheet – like many other European utilities – came under increasing pressure.

The company's three-year plan for 2012–14 foresaw investments of €2.6bn (US$3.35bn) in renewables, compared to €1.9bn (US$2.6bn) annually between 2009 and 2011. The company said it would add 1.45GW of new wind power during the period, compared to over 1GW per year in the three previous years. It also announced a major sale of 'non-core' wind assets and a big move into offshore wind.

Gamesa was a major casualty of the slowdown. The huge contracts it expected in Romania and elsewhere failed to materialise, and it realised it could not take Iberdrola for granted when it came to contracts. Gamesa was behind the game in offshore, and building new offshore turbine platforms was a capital-hungry affair at a time when its cash flow and debt figures were moving in the wrong direction.

As we have seen, Gamesa made aggressive moves to enter the Indian and Chinese markets, while trying to maintain sales in China through pursuing close relationships with the big utilities and by the end of the first quarter of 2011, 100 per cent of the company's sales came from outside Spain. The company moved to downsize much of its Spanish manufacturing. It also reduced capital expenditure on its planned 5MW and 7MW offshore turbine programme, moving a planned prototype of the first of its new machines from offshore in Virginia to a quay in the Canary Islands.

Success in emerging markets was not enough to stop Gamesa going into the red, and it reported its first quarterly loss in over a decade in March 2012, while its shares had fallen around 75 per cent over the previous year. Nor was it enough to save the job of Executive Chairman Jorge Calvet, who was replaced in May the same year, reportedly after pressure from Iberdrola. The new Executive Chairman was Ignacio Martin, yet another veteran of the auto industry – a former CEO of automobile components Cie Automotive SA (CIE) – who had been recruited to rationalise costs in the wind industry. In October 2012, Gamesa announced a major restructuring that would reduce Gamesa's work force by 20 per cent – a loss of 1,800 jobs – and align the company's operations with a new lower level of annual demand, by shutting down five manufacturing facilities, reducing capacity by 2GW. Financial analysts were impressed with the speed and aggressiveness with which Gamesa faced its problems, and compared the company favourably with Vestas' difficulties in restructuring.

By early 2013, the worst was over and, unlike Vestas, Gamesa was back in the black by the first quarter and for the first nine months it reported a net profit of

€31m (US$43m), turning around a €67m (US$93) loss at the same stage in 2012. Share prices had been recovering steadily from their 2012 low point. Martin told *Recharge* in Chicago in May:

> Previously Gamesa was focused on growth and expansion. Now the main focus is on performance. If the wind market is growing, that's excellent. But if there's no growth, then we still need to perform. We've reduced our break-even point ... we're launching new products and we're continuing with our geographical expansion so that we're never dependent on any specific market.
>
> (Stromsta 2013a)

The company's flagship 5MW upgrade to its revolutionary 4.5MW onshore turbine looked as if it was beginning to gain traction, with its first large-scale orders in Finland. It began energy production from its 5MW offshore prototype in the Canary Islands in October.

The company's challenges are far from over, however. Continued success in the markets where Gamesa has done well in recent years is far from guaranteed. As we have seen, India's market slowed down in 2012 and 2013, along with China. In Brazil, Gamesa has made an impressive grab for market share, but the new, more challenging FINAME regulations mean it will struggle to maintain profitable margins. And in the US, it is subject to the same stop-go market affects its peers and has found itself increasingly marginalised by a 'big three' of Vestas, GE and Siemens.

In offshore, it is arriving late to a market that already has a number of experienced incumbents and aggressive newcomers. It remains to be seen whether it can make an impact with a 5MW turbine, while plans for a 7MW machine are a long way from fruition. The long-standing relationship with Iberdrola is no guarantee of sales, as can be seen in the Spanish utility's decision to use Areva turbines in its 400MW Wikinger Project, and work with Siemens on its first UK project (which it is building with Danish utility DONG Energy). Like Vestas, Gamesa has a formidable accumulated knowledge of the wind market and highly regarded products. But like its larger Danish cousin, its balance sheet cannot compare to the diversified industrial companies, and it is 100 per cent dependent on the ups and downs of the global wind market.

Suzlon struggles with debt overhang

As we have seen, Indian turbine manufacturer Suzlon went through a meteoric expansion in the years leading up to the Copenhagen summit at the end of 2009, and – taking into account its REpower subsidiary – became the world's fourth-largest wind-turbine manufacturer.

The company's Chairman Tulsi Tanti had based his calculations on continued rapid growth of the global wind market. And while Suzlon was a leader in the onshore sector, offshore was taking much longer to get off the ground than Tanti had expected. A blade-cracking issue affected Suzlon's S88 wind turbine in 2008, forcing the company to pay around US$100m for a large-scale remediation programme.

The blade issue helped drag Suzlon into the red in 2009, and the losses continued in 2010, while the debt it had incurred to carry out and complete the REpower acquisition was becoming an ever more serious burden. Meanwhile, in China, Suzlon – in common with other non-Chinese players – found it was being squeezed out of the market by extremely low prices and margins, after having invested heavily in production facilities there. Suzlon's problems led to officials at rival Gamesa floating the idea of a merger or take-over – although it is not clear how seriously this was actually pursued, and Tanti was quick to stress that the company was not for sale.

Although the company had some successes, both in emerging markets like Brazil, and in offshore, its financial problems worsened throughout 2011–12. Although Suzlon's home market of India slowed as a whole in 2012, due to the phasing-out of Accelerated Depreciation, the most notable development is that Suzlon was not able to execute the orders that it had in its backlog because of restrictions on working capital. This led to Suzlon losing its leading position in the market in terms of annual installations for the first time to former Enercon subsidiary, Wind World (India). In a statement, Suzlon said 'Fiscal Year 2012–2013 for the Suzlon Group was focused on liability management, which resulted in a constrained business performance and lower than normal volumes, despite a strong order backlog' (Backwell 2013j).

Although Suzlon completed its 'squeeze out' of REpower in late 2011, attempts to use the profitable REpower division and its ability to raise finance to alleviate the parent company's balance sheet were unsuccessful, as REpower's German banks maintained a strict policy of ring-fencing. Despite announcing on several occasions that a more thorough integration of Suzlon and REpower was on the cards, observers were left somewhat bemused by the fact that no really major moves emerged, despite the prevailing overcapacity in the wind-turbine industry. Rumours of a 'reverse takeover' where the subsidiary would take over the parent company also turned out to be without substance, or the idea proved to be impractical.

Throughout the period, rumours circulated that Suzlon would be forced to sell off REpower to pay its Indian bank lenders and bondholders. In March 2012, a French engineering giant denied rumours that it was preparing a €1.5bn (US$2bn) bid to buy REpower. Officials from Suzlon also denied reports that it was trying to sell REpower, calling rumours 'completely speculative', although I was told by reliable sources that the company had opened a data room in Germany for potential investors through an investment bank.

The backdrop to the rumours was the looming issue of US$569m of payments on foreign-currency convertible bonds (FCCBs) later that year. In October 2012, Suzlon carried out India's biggest FCCB default – on US$221m of bonds – playing hardball with investors that refused to meet new terms, while continuing to work on a new deal with its syndicate of bank lenders.

According to investment bank HSBC 'Post the default on FCC repayments in October 2012 the company's operations (ex REpower) have come to a standstill, given the freeze on its bank accounts' (Charanjit Singh, HSBC, 24 June 2013, private correspondence). At the same time, Suzlon began a US$400m asset-sale programme, selling off – among other things – its China manufacturing business, which it had quietly shut down some time before.

In January 2013, it announced it had agreed a US$1.8bn package with a consortium of 19 banks to restructure its domestic debts under India's Corporate Debt Restructuring (CDR) mechanism, in a move that it said would help 'normalise' its business. The deal included a two-year moratorium on principal and term-debt interest payments, a 3 per cent reduction in interest rates and a six-month moratorium on working capital interest.

CFO Kirti Vagadia said the move was a 'major step forward in our efforts to achieve a sustainable capital structure' (Lee 2013b). At the time of writing, negotiations were continuing with bondholders. The CDR deal will probably allow Suzlon to reclaim its top spot in India in 2013, and the company should benefit from a speed-up in its home market from 2014 onwards, but Suzlon continues to be very cash-constrained. Sources in the company say that Indian lenders are pushing the company to sell off most of its non-Indian assets. Certainly, the evidence from once promising markets like Brazil suggests that Suzlon is carrying out a wholesale retreat from global markets.

Although officials deny that they are under any pressure from their bank lenders to sell REpower – renamed Senvion in November 2013 – the question of what do with the German subsidiary continued to be asked. Repower arguably had the potential to be more successful than its parent company, and analysts considered that the heavily indebted Suzlon was acting as a drag on the German company's prospects. At the same time, Suzlon is unable to fully integrate REpower without damaging or destroying it, while the debt overhang from the REpower acquisition was limiting the parent company's ability to operate successfully.

IHS Markit Director Eduard Sala de Vedruna said, 'REpower is vital to keeping the Suzlon Group afloat, but Suzlon's debt position inhibits REpower's capacity to finance new product developments or provision offshore warranties'. He warned that Suzlon needs to clarify 'the German vendor's future strategy inside or outside the Suzlon Group to avoid further value destruction' (*Recharge* 2013b).

Something had to give. Eventually, in 2015, Suzlon decided to sell Repower – now called Senvion, to US hedge fund Centerbridge for $1.2bn. This represented a substantial loss for Suzlon, but allowed it to overcome its debt problems and move on. We shall look at the prospects for the two companies in the next chapter.

Clipper – goodbye to the 'other' US OEM

The history of US turbine manufacturer Clipper, which enjoyed big success at the end of the last decade, can be traced back to the origins of the US wind industry and Zond. Clipper was formed in 2001 by Zond Energy founder Jim Dehlsen and his son, a few years after the Zond business had been sold first to Enron and then to GE.

Clipper expanded fast in the latter part of the decade, producing its 2.5MW Liberty turbine. Its US installations peaked at over 605MW in 2009, while it was also designing and producing the giant 10MW Britannia offshore turbine in the UK – mainly with UK government money – that was meant to spearhead the technology drive needed to develop the giant Round 3 zones.

While its technology was well regarded, and the Liberty was in some ways ahead of the game in a market dominated by GE's 1.5MW turbines, the company had several major weaknesses. First, like other OEMs it was subject to the US stop–go cycle and, unlike its rivals, it did not have a global manufacturing and market presence to compensate. Second, Clipper's growth was inordinately dependent on one company: oil major BP, which in the latter part of the decade was making a major commitment to wind energy. BP had backed Clipper's IPO in 2005, and the two companies were working together on the 5GW Titan project in South Dakota – one of the largest planned wind-farm complexes in the world.

Clipper was hit hard by a blade-cracking problem in 2007, which largely deprived it of customers outside of BP and hit its balance sheet hard, costing the company US$330m in a remediation programme to replace the blades on all its turbines. Then BP began to apply the brakes to its wind business growth and orders slumped, effectively sealing the fate of Clipper.

Clipper was teetering on the edge of bankruptcy by late 2009 after a loss of US$109m in the first half of the year, and put itself up for sale in September. Industrial conglomerate UTC – which produces among other things Sikorsky helicopters – completed an investment of US$207m to take a 49.5 per cent stake in Clipper in January 2011, and gained full control in for an additional US$223m in December that year.

At the time, Clipper CEO Douglas Pertz said that the sale to UTC was

> a transformational transaction for Clipper, bringing substantial capital from a strategic investor who is one of the world's leading industrial technology companies. Our relationship with UTC will enable Clipper to access UTC's support and expertise in areas of manufacturing, product quality and other industrial processes, while providing Clipper with equity financing to deliver our longer-term strategic goals.
>
> (Backwell 2009b)

Officials at UTC subsidiary, Pratt & Whitney Power Systems, said they were looking at plans to develop a 'next generation' wind turbine. In the event, UTC took a long hard look at Clipper and the wind-turbine industry and decided that it would be one guest too many at the party. UTC moved to cancel Britannia in August 2011. In August 2012, it sold Clipper to private equity group Platinum Equity for an unreported amount. At the time of writing, Platinum seemed focused mainly on trying to realise some value from maintaining Clipper's existing turbine fleet, and the company's production facilities were wound down in late 2012.

'UTC bought Clipper under certain market assumptions', says DaPrato, 'and most of them proved wrong. While the total investment was pocket change for a company of that size, UTC didn't want to wait five years to see if it could possibly pay off', he adds (Kessler 2012). 'We all make mistakes', Chief Financial Officer Greg Hayes said in March of UTC's brief flirtation with Clipper, after putting it up for sale (ibid.).

8 The wind industry bounces back

The difficult years of 2011–2013 were followed by a strong recovery in the 2014–2016 period, driven by a modest global economic recovery, the extraordinary boom in China's energy sector, steady expansion in core markets and the opening up of new frontier countries.

Recovery led to a new record year in 2015. The global industry installed almost 64GW, led by 31GW of installations in China. The US market installed 8.6 GW after a late surge, and Germany led a stronger than expected performance in Europe with a record 6 GW of new installations, including 2.3 GW offshore. Total global capacity reached 432GW at the end of 2015, representing cumulative growth of 17 per cent.

'Wind power is leading the charge in the transition away from fossil fuels', said Steve Sawyer, Secretary General of GWEC. 'Wind is blowing away the competition on price, performance and reliability, and we're seeing new markets open up across Africa, Asia and Latin America which will become the market leaders of the next decade', he added (Sawyer 2016).

By 2014, the market was booming and company balance sheets had recovered.

By 2015, company revenues, profit margins and valuations had reached new record levels. Perhaps the most remarkable story is that of Vestas, the Danish pioneer whose story we have followed across the chapters of this book. During

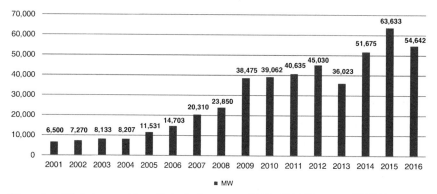

Figure 8.1 Global annual installed wind capacity 2001–2016 (Source: GWEC).

the slump, analysts and journalists were lining up to predict that Vestas would be bought out by a Chinese company or a major global industrial conglomerate. The line taken by commentators was that a 'pure play' wind-turbine manufacturer could not cope with the demands of a globalised market subject to strong fluctuations in regional and overall demand or compete with diversified competitors active across different technologies.

As we have seen, Anders Runevad took over the running of Vestas in 2013 and carried out a wholesale management shakeout, restructuring and cost out process, and invested in expanding profitable areas such as O&M and other services. The results have been nothing short of spectacular. Vestas has performed strongly in nearly all of the markets where it is present, re-entered markets where it had exited – such as India – and even managed to maintain a small but significant share of the market in China.

Vestas CEO Anders Runevad divides the company's journey from the low point of 2012 to renewed success into three phases: an operational phase; a growth phase; and a long-term phase. 'In the 2012 timeframe, Vestas was in a deep hole. Billion euro losses, our banks losing patience, the media criticizing everything we did or said', describes Runevad. 'Our customers were sticking with us, but it was tough. We were suffering from over-capacity, made worse by not having control over costs' (Runevad, personal communication).

Faced with this situation Vestas began the difficult process of cutting staff and production capacity to align the company to the realities of the market. 'In a sense, decision-making was fairly simple, because when you're losing that much money, every decision is geared to stopping the losses and starting to make money again', says Runevad. 'The overriding objective was simply to make sure the company was being properly run, costs were getting under control, and the bottom line starting to improve'.

In 2014, Vestas declared the turnaround a success and entered a new phase, which Runevad describes as a strategy based on 'growing profitably in new and emerging markets; building up the service business; and focusing on operational excellence'.

Today, Runevad says, Vestas is in a phase with a longer-term perspective. 'We are well-prepared as the market conditions evolve as they always do. We have a mid-term growth strategy that we update', he says, adding. 'The wind industry is moving from a period of solid growth to one of high and steady volumes, and we are preparing now for that future.'

US market shows Vestas' strength

One market in particular that shows the company's strength is the United States. There, the company has faced strong competition from US industrial giant GE – traditionally the largest seller in the US market – and Siemens. The triad of GE, Vestas and Siemens steadily increased their share of the market, with an array of other companies – most of them European – falling by the wayside or becoming marginal players.

While just a couple of years ago the company seemed to be under serious pressure in the US from Siemens in particular, the result of a period of intense competition between the three companies has seen Vestas emerge strongly supported by its growing industrial operations centred around a series of facilities in Pueblo, Colorado.

In 2016, Vestas supplied more turbines in the US than any other company – the first time that this has been the case, and was responsible for supplying fully 43 per cent of the 8.2GW of new capacity connected to the US power grid.

Globally, the story is similar. In 2014, some analysts put Siemens in first place in the global market while others continued to show Vestas in first place. In 2015, analysts put China's Goldwind in the top spot for the first time, supported by a by a booming domestic Chinese market. But in 2016, Vestas was back at number one in all of the analyst rankings, showing the resilience that a diversified presence across the globe provides it with.

Vestas has also made a series of aggressive moves to grow its service business. As the installed base of wind turbines grows, servicing the turbines is a continually growing business opportunity and Vestas has moved to build its 'multi-brand' servicing solutions – i.e., servicing non-Vestas turbines – partly through acquiring independent services companies UpWind Solutions (in the US market) and Availon Holding GMb. It has set itself a target of growing revenues from its service business organically by more than 50 per cent versus 2016 revenue.

Vestas' service activities have given the Danish company increased weight and allowed it to form new customer relationships, especially in the US. In May 2016, Vestas won a contract with Warren Buffet's Berkshire Hathaway Energy – which controls major utility MidAmerican Energy to service 1.75GW of GE turbines. It followed this in June with an agreement to supply 2GW of turbines to MidAmerican, a company it had not supplied before, beating both GE and Siemens, the incumbent suppliers (see Davidson 2016).

Siemens, which had been seeing Vestas grabbing market share from it in the US responded in November 2016 by teaming up with Duke Energy Services to provide multi-brand servicing. 'The agreement offers a one-stop shop for customers who are managing multiple brands of wind turbines in their fleet, helping them stay competitive and derive maximum value from their wind energy assets', said Duke Energy. '[It] means the market now has a new and powerful choice', the developer added (Weston 2016d).

Can it get any better?

Record valuations and revenues are the feature of today's wind-turbine OEM market. But this situation has not brought back the hubris and bloated company organisations of the noughties. Indeed, the processes set in motion by the last slump – consolidation and merger activity, restructuring, and globalisation seem to be accelerating rather than slowing down.

The reason for this is that companies are conscious that current market conditions are not likely to get much better, and nor are profit margins. Volumes of wind-turbine sales are likely to continue to increase steadily in the mid-term, but at lower prices. Competition is likely to intensify as the biggest players leverage their scale to increase market share. The long-awaited expansion of Chinese companies into world markets will materialise as the Chinese market reaches maturity and the key survivors emerge from its own consolidation, through a combination of market entry and acquisition.

'Despite booming installations, competition in the global wind market is more intense than ever, and Western turbine manufacturers are getting ready for the future by stepping up M&A activity', said Aris Karcanias, A Senior Managing Director at FTI Consulting and co-lead of the company's clean energy practice. 'Only those companies that can leverage global supply-chain economies and tap into high-growth markets around the world will be able to compete at the very top of the market.' (Karcanias, personal communication).

In short, all of the major turbine OEMs are looking over their shoulders at their competitors and trying to anticipate what their next moves will be. And some of them are carrying out pre-emptive moves, which they hope will keep them in the big leagues for the next decade or more.

At this point it would be useful to look at the main features of industry consolidation up until now, before asking 'what next?'

Consolidation arrives

Analysts and commentators have been predicting the consolidation of the wind-turbine industry, loosely meaning a reduction in the amount of manufacturing companies, and the emergence of a number of global market leaders for more than a decade.

As we have seen, two of the key features of the global wind industry have been; the relatively low level of automation in wind-turbine assembly, with most factories simply component assembly areas in warehouses; and the emergence of a large number of rival manufacturing companies, in large part due to the emergence of new national champions as new markets emerge.

As Alfonso Faubel, the former CEO of Alstom Wind described, most manufacturing facilities around the world are still mom and pop size operations compared to global manufacturing complexes like the automotive industry. While experts such as Henrik Stiesdal point out that the wind-turbine industry will never reach the per unit levels of production of the auto industry, it should reach a level approaching something like large truck production.

Historically industries have tended to move towards maturity as production reaches a certain level and markets become global. But the wind industry in the noughties seemed to be defying that trend, with the creation of new companies around the world – particularly in China, but even in the mature markets such as Germany. Consolidation, where it happened, was hesitant and limited.

Commentators waited in the vain for the kind of mega-deals that occur regularly in other industrial sectors.

All this has changed since the slowdown in growth and the crisis in company fortunes in the early part of this decade. We started this book with the announcement of the Vestas-Mitsubishi Heavy Industries merger for the offshore wind sector. But since then many more such deals have taken place as consolidation finally becomes a reality.

Key features

First, a number of players in mature and emerging markets have failed, stopped producing turbines, scaled back their ambitions, or been sold to larger rivals. Examples of the former include Fuhrlander, Prokon in Germany, M Torres in Spain and Clipper Windpower in the US. The Chinese market still has 22 companies producing wind-turbines, but the market is now highly concentrated with the big five accounting for 60 per cent of the total market. The market has also seen a number of Korean manufacturers who were poised to make major inroads in global markets – particularly in offshore – cancel or scale back their plans. These include the major chaebols Samsung, Doosan and Hyundai.

Second, the Tier One component suppliers – producers of blades and gearboxes for example – are also seeing a parallel process of consolidation. See for example, the acquisition of Hansen Gearboxes by Germany giant ZF, or the acquisition of blade manufacturing giant LM Windpower by GE.

Third, the offshore sector in particular – with its concentrated sales environment, giant size turbines, high R&D costs and strong balance sheet requirements, has driven the consolidation process and is likely to be dominated by just a handful of large players.

Fourth, the 'phoney' war between the biggest manufacturers – which prevailed during the slump years, with officials saying that they were reluctant to acquire further under-utilised capacity – is now over. In the course of a few years, some of the very largest industrial-scale wind-turbine manufacturers have been acquired or merged, and the global industrial conglomerates we introduced in Chapter 2 have well and truly taken the gloves off.

As FTI Consulting said in a report on the Nordex-Acciona Windpower merger: 'In a consolidating market – bigger and broader is better.' (Clover *et al.* 2015).

Consolidation timeline

The main M&A transactions until now are described in Figure 8.2.

MHI-Vestas

As we have seen, Vestas and Mitsubishi Heavy Industry created a new company, MHI-Vestas Offshore Wind, in order to share manufacturing investments and compete in the fast growing offshore wind segment.

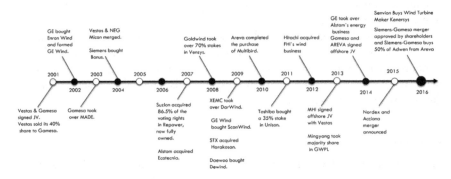

Figure 8.2 Consolidation timeline (Source: FTI Consulting).

The joint venture allowed the completion of the 8MW V164 turbine, which became the largest turbine in the market when the first prototype was deployed in January 2014. In 2015 the company received its first order for the new turbine, from DONG Energy for the Burbo Bank Extension project and since then has received a series of major orders. While the latest figures from industry association still show Siemens as still the undisputed leader in terms of offshore installations, these figures will change over the coming years as installations and production from MHI-Vestas ramp up. And just as importantly, the merger allowed Vestas to reduce its Capex spending at a time when it was in a period of critical strain and needed to restructure.

Hansen-ZF-Bosch Rexroth

In 2011, Germany engineering and automotive component manufacturer ZF Friedrichshafen acquired Belgium-based Hansen Transmissions, one of the largest producers of wind-turbine gearboxes from Indian turbine manufacturer Suzlon, and other investors for around US$520m. The combined company has the capacity to supply 11GW of wind-turbine gearboxes per year and competes in the global market with Siemens-owned gearbox supplier Winergy (with around 10–12GW of capacity per year) and China's High Speed Transmission Equipment Group, which at the time had similar capacity. Notably, the combined capacity of the three giants in the 2012 period was roughly equal to the total amount of installations of geared turbines that same year. There was also significant production from smaller but well-placed players such as Bosch Rexroth, Moventas, Eickhoff and Chinese players such as Chongqing group and DHI, pointing to continuing over-capacity in the market. In 2015, ZF acquired the wind gearbox operations of Bosch-Rexroth for an undisclosed price, acquiring production lines in Germany and China and further consolidating the market.

Gamesa-Areva-Adwen

As we have seen, Spanish 'pure play' wind-turbine manufacturer Gamesa was hit by the same headwinds that other manufacturers such as Vestas suffered in the early part of this decade. Gamesa's problems were made even more extreme by the fact that demand in its home market in Spain had dried up after a dramatic political policy U-turn. Owners including Iberdrola demanded a wave of cost cutting and restructuring, which left it poorly placed to complete development of new 5MW and 7MW offshore turbine models.

At the same Areva Wind, in a move – uniquely – to create a pure-play *offshore* wind company, was struggling to ramp up operations for its existing 5MW Multibrid turbine and develop a new 7–8MW machine that could compete with the likes of Siemens and MHI-Vestas. Crucially, Areva Wind's owner, state-controlled French nuclear firm Areva, was in serious problems having incurred huge losses from the development of its new reactor design, and was in the process of being broken up. Areva officials had made it clear that there was simply no more cash for the loss-leading Areva Wind operation.

In 2014, Gamesa and Areva Wind announced plans to work together in off-shore, and after extended negotiations, the two companies announced the formation of Adwen, a 50:50 JV, in April 2015. Adwen, the two companies announced, would jointly commercialise existing 5MW turbine models and develop the 8MW model, mainly based on Areva Wind's existing design.

The two companies said the new company had a sales pipeline of 2.8GW of projects under development and a goal of controlling 20 per cent of the European market by 2020. While there was widespread excitement in the market about the capabilities of the yet-to-be-built 8MW, there were serious doubts about the combined company's ability to bring the turbine successfully to market, given the continued doubts around the Areva's future and the fast-moving progress of the joint-venture's competitors. There were continued rumours of a sale of Adwen to other market participants.

GE-Alstom

The first major consolidation of the market through acquisition was the high-profile purchase of France's Alstom Power by the US company GE in a $10bn deal in 2014–15. The deal encompassed Alstom's wide range of power and grid assets and allowed Alstom to keep its rail division, while forcing the divestment of its gas turbine division to Italy's Ansaldo. Alstom's Renewable Power business alone had close to 10,000 employees and sales of around €2 billion in 2013/14.

The deal involved negotiation at the highest level with France's President and his officials. A key role was reportedly played by then Deputy Economy Minister – and now President – Emmanuelle Macron, who reportedly convinced his boss, Economy Minister Arnaud Montebourg to desist with his plan of blocking the move in favour of a European merger with Germany engineering giant Siemens. Macron's intervention allowed the deal to be finally approved by Alstom's board in June 2015, before receiving approval from EU and US regulators in September.

As part of the quid pro quos involved in the deal, GE agreed to spin off its renewable energy assets into a new standalone division, GE Renewable Energy, which would be relocated from its traditional Schenectady, New York, base to Paris. The CEO of the new division would be a French former Alstom official Jérôme Pécresse, who now reports directly to GE CEO Jeff Immelt.

In onshore wind, Alstom was a 2nd tier, but significant player, with sales mainly in Europe and Brazil, where it had become the biggest player in the market and established two turbine manufacturing complexes. Coincidentally, GE was in second place in the Brazilian market, giving the combined company a market share twice as big as its closest competitor, Gamesa.

No pain, no gain in onshore

The GE takeover has not been without pain for Alstom's existing operations in onshore wind. In January 2016, GE announced it was planning to cut 6,500 jobs across the Alstom energy businesses in Europe over the next two years. After a thorough internal review, GE moved to kill off Alstom's product lines in most markets in 2016.

In Brazil, where the two companies were the market leaders, the situation is particularly painful, and has been made worse by the slowdown in the country's burgeoning wind market in the wake of its political and economic crisis (see Chapter 4).

As we have seen, Brazil's wind-power installations fell from 2,754GW in 2015 to 2,014MW in 2016, and more importantly the pipeline of future orders dried up amid the cancellation of planned power tenders. 'This year we've had no volumes allocated', said GE Renewable Energy CEO Jérôme Pécresse in June 2016. 'We have no visibility on future volumes to be contracted and there are very important delays in payments from customers.' He went on to warn that 'there is a risk of the supply chain collapsing' in Brazil and noted that GE has reduced its workforce and has idle capacity at its three wind-equipment manufacturing plants in Brazil (Dezem 2016).

Alstom deal revives GE's offshore plans

In offshore, GE decided to cancel its offshore plans around its 4.1MW turbine – the result of an acquisition of Norwegian company Scanwind – in 2013. Alstom, in contrast, had created a strong position in the market, helped by French government tendering practices that saw it commit to creating a major new manufacturing complex for offshore wind turbines. However, there had been persistent doubts among analysts around Alstom's 6MW Haliade turbine's ability to compete in a global market that was soon to be dominated by machines in the

8–10MW range, as well as persistent reports around mechanical teething prob-lems. Any deal between GE and France was bound to have included strong commitment to France's offshore wind industrialisation plans. GE seems to have picked up the Haliade programme and run with it. The first fully fledged Haliade project was completed at Block Island, off the coast of Massachusetts in May 2017. And GE officials have since said that they are working on a new, 10MW turbine, indicating that GE is back in the offshore game for good.

GE Blade Dynamics, GE-LM Wind

A year later, GE followed its acquisition of Alstom by buying the world's biggest independent wind-turbine blade manufacturer LM Windpower from private equity player Doughty Hanson for € 1.5bn.

As well as supplying (along with others) blades to GE, LM is a major supplier to Spain's Gamesa and China's Goldwind. EU regulators, approving the deal on 20 March 2017, brushed off concerns around any unfair advantage the deal could give GE over its competitors, concluding that 'competing blade manufacturers would continue to have access to wind-turbine manufacturers other than GE'. It added that 'GE would continue to face significant competition from other major turbine manufacturers, such as Siemens, (MHI) Vestas, Nordex and Senvion, who either manufacture their blades in-house and/or are not dependent on LM Wind Power for supplies.' (Toplensky 2017). GE had also made a smaller but significant statement of intent a few months earlier when it bought innovative modular blade design and manufacturing Blade Dynamics in October 2015 for an undisclosed sum.

Suzlon-Senvion

While GE, Alstom, Gamesa and others were stepping up M&A activity, Indian wind-turbine manufacturer Senvion was planning a de-merger. Senvion, who we first met in Chapter 4, had grown – in large part through an acquisition strategy – into a global giant that had become the fourth-largest ranking manufacturer at its high point and controlled German offshore and onshore manufacturer REpower (its name was changed to Senvion due to a trademark clash in 2014).

The temporary slowdown in the Indian and global market, Suzlon's growing debt overhang and ring-fencing covenants involving Senvion's German creditors, meant that it was simply too big to hang onto. Following the sale of Hansen in 2011, and after a series of attempts to either launch an IPO or sell Senvion to different parties, it was sold to US head fund Centerbridge in January 2015 for $1.2bn. The sale allowed Suzlon to reduce its debt burden significantly and concentrate on rebuilding its hitherto dominant position in the now accelerating Indian market.

For Senvion, still a significant player in the global market, the path was less clear. Escaping from the clutches of Suzlon means that Senvion is now free to compete in a number of fast growing markets where it was previously unable to do so. Centrebridge's ownership should also – at least on paper – give it the resources to increase spending on R&D to upgrade both offshore and onshore turbine models. In 2016, the company announced several of its own acquisitions, including Indian turbine manufacturer Kenersys (meaning it is now competing on its former owner's doorstep) and blade developer EUROS.

Nordex-Acciona Windpower

Next on the merger trail were German turbine manufacturer Nordex and Acciona Windpower. Nordex is a high-quality producer with an important share of its domestic market and a significant niche presence in a number of European and non-European onshore markets.

Acciona-Windpower was originally a mainly in-house turbine producer for Spanish renewables giant Acciona Energy, which had expanded its international sales to third parties through the introduction of a successful 3MW turbine platform using high altitude concrete towers.

Nordex and Acciona announced that they were merging in October 2015, in a cash-plus-shares deal worth around €785m. The deal gained regulatory approval in April the following the year.

Under the terms of the deal, Nordex acquired Acciona Windpower from its parent company, energy and infrastructure conglomerate Acciona Group. However, Acciona Group acquired a 29.9 per cent stake in Nordex, making it the company's largest shareholder. Together, the two companies had sales of €3.4bn in 2015 and had installed turbine capacity of 18GW around the world.

'We are now taking the first steps towards establishing our company as a truly global player in the wind turbine industry', said Lars Bondo Krogsgaard, CEO of Nordex (Weston 2016c). On the face of it, the deal looks to have compelling strategic logic, and is 'completely additive' to both companies in terms of product range, technology, client types and geographies, according to FTI Consulting. The two companies expect to achieve around €95 million of synergies by 2019. Around half of this will come from sales synergies with 40 per cent from supply chain/procurement synergies and the balance from integrated R&D.

There is a good geographical fit between the businesses, with Acciona Windpower effectively bringing Nordex a more global presence and emerging market exposure. In particular, Acciona Windpower will enhance Nordex's presence in fast-growing emerging markets such as South Africa, Brazil and Mexico, and also the US, away from its core business in mature European markets.

Around 80 per cent of Nordex's order intake comes from the EMEA region. In contrast, the Americas currently represent 94 per cent of Acciona Windpower's orders, roughly shared between Latin America, the US and Canada. Acciona has turbine assembly plants in Spain, the US and Brazil, and is currently constructing a facility in India. It is only in the US and South Africa markets where there is significant overlap in sales between the two companies.

From a technology perspective, the deal also seemed to make sense. Nordex has made low- to mid-wind-speed (IEC II and IEC III) sites a sales strength while Acciona Windpower's product offering is concentrated in higher wind speeds, so the combined offering will have more strength across all wind speeds. In addition, Nordex expects to benefit from Acciona Windpower's expertise in using cheaper concrete towers and Acciona Windpower will benefit from Nordex's expertise in low-noise blades using carbon fibre with integrated anti-icing capability. In practice it remains to be seen how easily these blade and tower technologies can be cross-pollinated and whether expected levelised cost of energy (LCOE) savings can be achieved.

From a client perspective Nordex's focus has been on small- to mid-size customers with projects on average below 30MW and often close to densely populated areas. Acciona Windpower's focus has been mainly on utilities and IPPs and on plants of 100MW and above, mainly in remote areas that have no land constraint issues. In addition, Nordex will benefit from a new strategic relationship with global developer Acciona Energia as a client. It appears that the new combined turbine manufacturer will get right of first refusal on all Acciona Energia projects. Acciona Energia was the seventh biggest wind asset owner as of the end of 2014 (with around 7GW of net owned wind assets), based on net ownership although growth has slowed recently.

At the time of the merger, the two companies had the stated strategic aim of achieving a top five spot in the global wind-turbine OEM rankings, with the combined entity expecting by 2017 to capture 8–10 per cent of the global market, compared to a combined share of 4.2 per cent in 2014. Given the moves being made by Nordex-Acciona's competitors and the more intense competition in the market as growth flattens out, the target looks over ambitious.

In 2016, the combined company had a global market share of 4.8 per cent, putting it in seventh place, according to FTI Consulting. However, the new Nordex may be able to make significant gains in coming years further on if it can successfully lever the companies' combined strength.

As with all M&A, the integration is not without its risks. In particular, Nordex the management of Acciona are having to navigate differences in management culture and style. And it will also have to balance the interests of the turbine manufacturer and the wider Acciona Group, which now becomes the anchor shareholder of Nordex.

Up until the time of writing, it has not been all plain sailing. The deal initially put the existing Nordex management in the driving seat, with former CEO Lars Bondo Krogsgaard – a veteran of DONG Energy and Siemens – staying in his role. In March 2017, however, Nordex Acciona issued a profit warning, announced lower than expected revenue forecasts and warned of increased 'price pressure' in the market. The company's shares fell around 35 per cent after it announced revenues in 2017 should be €3.1–€3.3bn, and €3.4–3.6bn in 2018, down from an expected €4.2bn (Weston 2017b).

Krogsgaard resigned from the company after the profit warning, and was replaced by former Acciona Windpower CEO and Nordex-Acciona COO and deputy CEO, José Luis Blanco. On his resignation, Krogsgaard said: 'Nordex is fundamentally stronger than ever, but our credibility has suffered as a consequence of the outlook, and this follows a period, where our handling of communication matters has been criticised. This is, ultimately, my responsibility.'

Nordex chairman of the supervisory board Wolfgang Ziebart said: 'During Lars' tenure our company has more than tripled in size at steadily improving margins, and we have become a truly global player. I regret Lars' departure, but the full supervisory board respects his decision to do what is considered to be in the best interest of the company. In appointing José Luis Blanco as CEO we ensure the continuity of the company's performance.' (Weston 2017b).

Analysts welcomed the change. 'As an executive focused on operations, we see Blanco as well positioned to manage the challenges of integrating Nordex with Acciona Windpower', said investment Bank HSBC. 'His appointment will reinforce the influence of Acciona in Nordex', it added (Vidal 2017).

Siemens-Gamesa

But the biggest merger of all was still to occur. In January 2016, Siemens Wind and Gamesa announced that they were in talks to discuss a merger, following persistent rumours that a deal was being considered. After six months of negotiations, the companies announced that an agreement had been signed in June and, received regulatory approval for the merger from the EU in March 2017.

Under the terms of the deal, Siemens, agreed to pay €3.75 a share in cash to Gamesa shareholders, taking control of a 59 per cent stake in the new company, which would be delisted and headquartered in Spain.

By carrying out the merger, Siemens and Gamesa have created the world's largest turbine company, valued at around €10bn. Siemens Gamesa Windpower was expected to have an order backlog of around €20bn, annual revenues of €9.3bn and operating profit of €839m. Ignacio Martín, executive chairman of Gamesa, said: 'The merger with Siemens constitutes recognition for the work performed by the company in recent years and evidences our commitment to generating value in the long term by creating significant synergies and extending the horizon of our profitable growth.' He predicted the combined group would become the dominant player in the wind-turbine market. 'Today, we are embarking on a new era, creating, alongside Siemens, a world-leading wind player', he told investors (*Financial Times* 2016).

The deal aimed to create a market-leading company that is positioned to resist increased competition from the other giants created by the merger process and burgeoning Chinese companies. In 2015, before the deal, Siemens had a 9.5 per cent share of the global wind-turbine market, according to data from FTI Consulting, compared to Vestas, which had a share of nearly 12 per cent. Gamesa had a 4.5 per cent share of the market.

On paper, the two companies are a good fit. '[The merger] opens up a number of the important, emerging and established markets for Siemens', says Aris Karcanias, a Senior Managing Director and wind industry expert at FTI

Consulting. 'Gamesa also has a strong and flexible supply chain and has shown itself adept at bringing new turbine offerings to the market at competitive cost', he adds, 'allowing it to offer a lower cost of energy as well as a stronger service proposition' (Weston 2016e).

Both the GE-Alstom and Nordex-Acciona Windpower mergers had been set to increase competitive pressure in the onshore market. And the key challenge for Siemens had been to gain traction in fast growing emerging markets.

FTI Intelligence forecasts that globally the wind market will grow at a compound annual growth rate (CAGR) of 3.3 per cent 2014–2024, but many emerging markets are forecast by FTI Intelligence to grow at over 25 per cent per annum to 2020. While Siemens has performed strongly in the United States, Europe and in the offshore sector where it is dominant, breaking into emerging markets has been more challenging. Gamesa, on the other hand, is the number one vendor in India and Mexico, and number two in Brazil, whereas Siemens has struggled to make an impact in all three markets.

Market and regulatory developments also drove the merger. As we have seen, governments have increasingly moved away from Feed-in Tariff schemes towards tender-based mechanisms such as long-term capacity auctions. These mechanisms create much stronger pressures for projects to be competitive in cost-of-energy terms. Siemens' product portfolio has traditionally been focused on reliability and quality. Gamesa, on the other hand, is well known for its extremely lean manufacturing and ability to produce good technology at a lower cost, for example in Brazil whose auction system has put intense pressure on margins.

Siemens can learn some tricks from Gamesa in service market and technology areas. Services and O&M are becoming an ever more important part of securing companies' revenues as installed capacity grows and turbines come out of their warranty periods and age. Gamesa leads Siemens on flexibility and scope, the breadth of its operating and its ability to service other manufacturers' turbines – so-called multi-vendor servicing.

In terms of technology, the merger is not exactly a match made in heaven, however. Siemens has put an increasing emphasis on direct-drive technology – both offshore and onshore, whereas Gamesa has taken a medium-speed gearbox approach.

However the benefit for Siemens is that Gamesa has moved faster and more decisively than Siemens to develop products for the strategically important low-wind segment, and has deep experience in designing turbines for areas with weak grids. Siemens will benefit from Gamesa's experience in squeezing out costs, something which it has done particularly successfully over the last five years.

Analysts point out that as diversified supplier to the energy market which is exposed to the oil and gas sector, the Gamesa takeover is a clear sign that the Siemens board sees the strategic importance of wind power in the energy market and maybe looking at a shift of its strategic focus in the energy market as a whole.

Finally, there are few pure-play wind-turbine manufacturers left for large industrials to acquire, so Gamesa represented one of the last remaining chances for a company like Siemens to grow in scale and ensure it can continue to compete with Vestas, GE and Goldwind. The other big pure-play survivor, Vestas, with a market capitalisation of around € 13.5bn ahead of the merger may have been too big a bite for Siemens to take, while EU competition authorities were likely also to have looked unkindly upon any such development.

Absorbing Gamesa will not be straightforward from an industrial and technology standpoint. The two companies have had very different technology and manufacturing philosophies: Siemens' traditional focus on robustness and quality versus Gamesa's focus on low-cost R&D and manufacturing. Integrating the two product portfolios will be a big intellectual, cultural and operational challenge.

Having said all this, Siemens has much to gain from making a decisive move at this stage. Acquiring Gamesa and integrating the company successfully should allow it to address some of the biggest challenges its wind business currently faces, in terms of its turbine portfolio and market exposure (Karcanias *et al.* 2016). Ultimately, making a move for Gamesa has propelled Siemens to the top of the global turbine market, and will prepare it for survival in the face of competition to come.

Before and after … creating a European giant

The new company's management was announced in May 2017. Marcus Tacke, the former CEO of Siemens Wind, was named as CEO of the new combined company, while Siemens' Michael Hannibal was named CEO of the company's offshore division. Other members of the company's senior management team

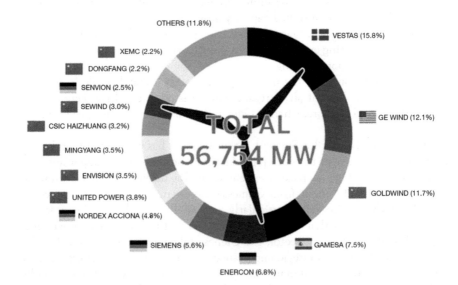

Figure 8.3 Top 15 wind-turbine suppliers in the annual global market in 2016, when Siemens and Gamesa were separate companies (Source: FTI Intelligence).

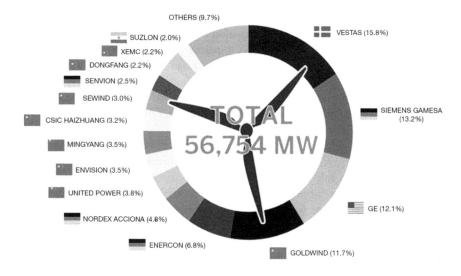

Figure 8.4 Top 15 wind-turbine suppliers in the annual global market in 2016 after the merger of Siemens and Gamesa (Source: FTI Intelligence).

include: Andrew Hall, chief financial officer; Xabier Etxeberria, CEO onshore; Mark Albenze, CEO service; and David Mesonero, corporate development, strategy and integration managing director.

'We are already a leader in the renewable energy marketplace', said Tacke, 'but we have much work ahead of us. Our priorities are clear: delivering on our projects, winning new business and creating a company culture focused on engineering excellence and vigorous cost management.' (Siemens Gamesa 2017).

What now for offshore?

As we have seen Siemens has been the market leader in offshore, while Gamesa had formed the Adwen joint venture with Areva precisely in order to challenge Siemens and the other leading companies in the sector. As Siemens and Gamesa began to negotiate their merger, there were concerns that Siemens' takeover could create anti-trust concerns among European regulators because of its dominant position in offshore, and it was also not clear how Adwen would fit in to a merged company.

There were persistent market rumours that Adwen could be purchased by GE, which would see the US company develop Adwen's 8MW turbine platform as a replacement for the Alstom-developed Haliade machine, or by rival offshore producer Senvion. However, Gamesa reportedly rejected a GE offer for some of Adwen's assets, and exercised its option to purchase Areva's stake in the joint venture for €60m in September 2015, finally ending the French nuclear company's foray into offshore wind.

FTI Consulting's Feng Zhao said Gamesa's decision to hold on to Adwen will end up enhancing Siemens' leading position in the offshore wind sector. 'Benefits include an approximate 2GW project pipeline and the addition of one more offshore turbine drive-train solution', says Zhao (personal communcation). However he notes that there are still concerns about how Siemens will deal with the issue of two factories (for nacelles and blades) in France that Adwen had agreed to build, while in Germany, both Adwen and Siemens have facilities in the town of Bremerhaven.

As the company currently stands, the merged Siemens-Gamesa OEM will own two 8MW turbines, with different gearbox configurations. Siemens' SWT-8.0-154 was revealed in July 2016, as an evolution of its 6MW then 7MW turbines, with testing and type certification is expected in 2017. Adwen's 8MW, with a world-leading 180-metre rotor, is also set to enter the testing phase, after the prototype was successfully erected in May 2017. This leaves Siemens-Gamesa with some interesting choices to make about which technology paths and production models it wants to follow.

The effect of consolidation for the offshore sector is striking. The Siemens-Gamesa-Adwen deals leaves just two offshore manufacturers with the new market-standard 8MW turbine on offer. This compared with six companies that offered 5–6MW turbines just a few years ago.

A new era of competition

To summarize, the wind market returned to boom in 2014–2015, and share prices for the major companies in the sector reached levels not seen since Q3 2008 before the Lehman Brothers-induced financial collapse. EBIT margins for the sector headed back towards double-digit levels and balance sheets were significantly repaired.

The industry aggressively moved to address its cost base after the difficult years of 2011–12 and this resulted not only in stronger companies, but in significant advances in the levelised cost of energy (LCOE) and a more sustainable long-term growth outlook for the sector (see Chapter 9).

However, following the China boom of 2015, the sector grew more slowly in 2016. 'Wind power continues to grow in double digits; but we can't expect the industry to set a new record every single year', said Steve Sawyer, GWEC Secretary General. 'The industry is set for a levelling off in growth during 2017–2020, with annual installations continuing at a healthy but slower rate, before faster growth returns in the early 2020s.'

Global wind market forecast 2015–2024

The new market situation is one of higher volumes but lower prices, and turbine manufacturers have spent the last few years consolidating, trying to find complementarities, and creating greater economies of scale. Against a backdrop of slower growth, many turbine manufacturers have said they expect to grow 'faster than the

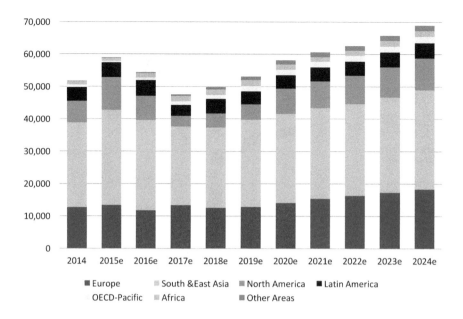

Figure 8.5 Global wind market forecast 2015–2024 (Source: FTI Intelligence).

market', with Nordex-Acciona for instance suggesting that the combined entity would grow 30 per cent in sales terms by 2018 from a 2015 base. Most of the big companies expect margins to be either flat or to improve slightly. The question is whether continued growth and margin expansion is possible when margins are close to peak and the annual market is set to decline in the short term.

Chinese turbine exports are expected to increase slowly but steadily, given a greater presence in project development and capacity tenders. These latter, will continue to drive more competitive turbine pricing.

The Siemens-Gamesa merger has created a European wind-turbine mega-company that is set – on paper – to be the market leader in the coming period.

However, the recent performance of Vestas – particularly in the US where competition is fierce and conditions similar for the top companies – suggests that it is in a strong position to keep challenging for the number one spot. Vestas CEO Anders Runevad sees his company as being well-placed to continue to lead in the new world of price competition and big consolidated players:

> There is a relentless demand for lowering the cost of energy and with market-based policies like auctions being 'the new norm', it is more import-ant than ever that we sharpen our focus and continually evolve to meet the competitive environment – both from other OEMs that are increasingly consolidating as well as from other power generation sources.
>
> (Runevad, personal communication)

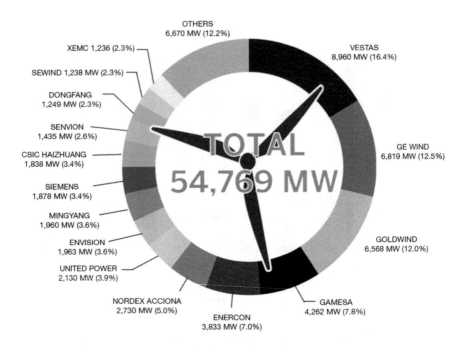

OTHERS
6,670 MW (12.2%)

XEMC 1,236 (2.3%)

SEWIND 1,238 MW (2.3%)

DONGFANG
1,249 MW (2.3%)

SENVION
1,435 MW (2.6%)

CSIC HAIZHUANG
1,838 MW (3.4%)

SIEMENS
1,878 MW (3.4%)

MINGYANG
1,960 MW (3.6%)

ENVISION
1,963 MW (3.6%)

UNITED POWER
2,130 MW (3.9%)

NORDEX ACCIONA
2,730 MW (5.0%)

ENERCON
3,833 MW (7.0%)

GAMESA
4,262 MW (7.8%)

GOLDWIND
6,568 MW (12.0%)

GE WIND
6,819 MW (12.5%)

VESTAS
8,960 MW (16.4%)

TOTAL
54,769 MW

Figure 8.6　Annual onshore wind-turbine suppliers in 2016 (Source: FTI Intelligence).

Although Vestas sees changes to the market environment and industry consolidation as a healthy development as the wind sector matures, Runevad says that this means Vestas needs to act on multiple fronts to maintain its market leading position, strategically and organisationally.

Vestas points out that the size of its installed base of 82GW of turbines – in 76 countries – its backlog, and the number of turbines under Vestas' service are unparalleled. And together with its accumulated know-how and truly global set-up, we can expect Vestas to be leading the development of the wind-turbine market for the foreseeable future.

The leading Chinese companies meanwhile, are likely to be outside of the top three places for the next few years as China's market pauses for breath ahead of the next big push. But they will also use this period to continue growing their market shares and their operational footprint across global markets.

9 Tipping point

Windpower's iPhone moment

In August 2016, the small but dynamic Latin American country of Chile held a huge power tender that would define the future direction of its energy. The tender, equivalent to around 1GW of power capacity, allowed all types of generator – fossil fuel or renewable, new projects or existing plants – to compete to supply 'firm' power for delivery in 2020–21. To the surprise of many, renewable power producers, led by Italy's Enel and prolific Irish independent Mainstream Renewable Power won the lion's share of the capacity on offer, at prices that averaged US$40/MWh for wind. Incumbent generators that had dominated the Chilean market for decades were left rubbing their wounds. The future of Chile's power system, it seems, is renewable power.

As we have seen, the story of wind power since the late 1970s has been one of almost unmatched growth, punctuated by periods of slowdown, often sparking crisis for the companies involved. We have seen that the main cause of these slumps has been changes in government policy that affect important national markets. As one market moves into a period of boom, others have seen governments abruptly choke development, disrupting supply chains and hitting companies over-exposed to particular markets.

While the wind industry is exposed to changes in macroeconomic conditions like other industries – both globally and nationally, I would argue that what has given it a unique vulnerability is the fact that wind power has almost always depended on government support and/or subsidies to grow. This support is motivated by countries' necessity to reduce carbon emissions, either individually or as a part of international climate agreements, and to grow clean-energy industries to scale in order to do this. These political arrangements have shown themselves highly susceptible to disruption at times of economic stress as politicians use the arguments of incumbent fossil fuel lobbies and climate sceptics to justify slowing momentum, or simply argue that the public purse cannot support further funding.

As a strategic aim, the wind industry and other renewables sectors have long put forward the need to reach 'grid parity' – in other words, to be able supply electricity into the grid at an equal or lower price to competing power sources such as fossil and nuclear. In reality, the world's energy system is characterised by complex, overlapping layers of regulation and subsidy – and as many have pointed out, most of this goes to the established fossil and nuclear industries.

But the good news is that grid parity has now been reached in many of the world's power market, at least for new power capacity. Renewables costs have been falling steadily behind the scenes, in prolific, well-established regions such as the Texas Panhandle.

According to the 10th annual Levelised Cost of Energy Survey by financial advisory and asset management firm Lazard, today's average wind costs are one-third what they were in 2009, having fallen from US$140 per megawatt-hour to US$47 per megawatt-hour in just seven years and making it the cheapest available form of power in many markets, even compared to already constructed fossil fuel plants (see Lazard 2016).

The new-found efficiency and cost structure of the global industry has only just started to be understood and filter through into mainstream media comment and policy making, as a series of highly visible public tenders has illustrated the new state of play.

Some of the highlights so far have been:

- **Brazil**, where a tender in 2012 saw prices as low as US$42.20/MWh driven by a very competitive tender framework and industry beating capacity factors.
- **South Africa**, where companies' bid prices in the fourth Renewable Energy Independent Power Procurement Programme (REIPPPP) tender at R560/MWh (US$47.14) around half of prices for new coal-fired plants.
- **Morocco**, which established a new global low for wind-power projects when it announced the results of a 850MW tender, with winning prices between US$25/MWh and US$30/MWh, and wining companies including Italy's Enel Green Power, France's EDF Energy, Spain's Acciona and Saudi Arabia's ACWA Power. The country's vice minister for energy and environment, Abderrahim El Hafidi said, 'Things have changed a lot. ... Now, we have wind projects cheaper than coal. The US$30/MWh bid compares to coal which is US$80/MWh.' He added: 'Isn't that amazing that we can have confidence in renewable energy for the future of our energy and for the future of the planet. This is real. It is not a claim.' (Parkinson 2016).
- **Mexico**, where the country's second power tender saw average wind prices at US$32/MWh. Tender prices have also driven very competitive deals for long-term wind Power Purchase Agreements (PPAs) in Mexico's private market.
- **Peru**, whose renewable energy tender in February 2016 saw successful bids for three wind farms with a capacity of 162MW awarded at an average price of US$37.79/MWh, 'the lowest price awarded to wind projects in Latin America in recent years', as energy and mining regulator Osinergmin noted. Enel Green Power, the 'eight hundred pound gorilla' in many of the recent developing market tenders won three-quarters of the capacity on offer.
- **Chile**, where, as we have seen, wind prices in the August 2016 tender averaged US$40/MWh, compared to energy prices at previous tenders of around US$100/MWh.
- **Argentina**, which has organised Round 1 and Round 1.5 tenders to date (with a second tender underway as this book was completed). Wind prices in

the tender 1 and 1.5 averaged US$59.4/MWh andUS$54/MWh respectively, with the lowest wind price in 1.5 at US$46/MWh, despite the fact that only a small amount of wind-power capacity had been built in Argentina up until the time of the tender.

- **European offshore wind**, where a series of tenders in 2016 and 2017 saw very rapid falls in European offshore wind prices. Prices were €54.5/MWh in the Netherlands and €49.9/MWh Denmark, while in Germany's 2017 offshore wind auction two companies, DONG and EnBW, made bids at effectively the same level as wholesale electricity prices, which are currently around €30/Mh.

There are also a number of onshore tenders currently underway in European countries such as Germany and Italy that have recently made a transition to competitive auctioning, which are leading to steady price falls. We can expect a number of other countries, including the UK to show very competitive onshore wind prices once they have provided a framework for continued tendering.

The impact of competitive wind – along with even more spectacular falls in the costs of wind's sister renewable technology, solar PV – has been enormous. Previously sceptical editors at major media outlets such as the *Financial Times*, the *Wall Street Journal* and the *Economist* have sat up and taken notice of the new environment, along with existing media champions such as Bloomberg. The new economics are filtering fast into policy-making circles around the world, helped by the work of IRENA and the previous fossil-centric IEA.

Wind industry briefings with politicians and government officials had previously taken the form of emphasising climate targets and showing future projections for renewables costs. Now, briefings are a matter of cold, hard numbers from recent tenders and the simple message: 'We are cheaper than they are.'

The slogan used by Wind Energy Europe (WE Europe) for its intervention at the Paris climate talks in 2015, 'Clean, competitive, ready', summed up the new mood in the industry, with WE CEO Giles Dickson telling an OECD event during the talks 'we need to move to wind energy because it simply makes economic sense.'

The new rules for competitive wind

As we have seen, the wind industry has entered into a new stage of its development, with something like grid parity being established in markets around the world. But which factors, both external and from within the industry itself – have driven the new competitiveness and allowed the industry – in the words of Mainstream Renewable Power founder and CEO Eddie O'Connor to reach its 'iPhone moment'? (See O'Connor 2017).

Shale gas

By the end of the 'noughties' it was clear that changes in oil and gas technology – which permitted the practice known as 'fracking' and the industry that grew up around this – were creating major and perhaps irreversible changes in the dynamics of the oil and gas business.

Environmentally destructive and controversial, the shale industry in the US has been able to bring vast amounts of untapped hydrocarbon resources on line. While still more expensive than conventional production in the best oil provinces, the shale industry's key feature is its ability to survive and thrive through a series of boom–bust cycles and respond to small changes in oil and gas prices, scaling up or down production risks and labour at incredible speed, and accessing an eco-system of investors and finance.

Whereas throughout most of the last decades of the twentieth century, the US economy was vulnerable to international oil prices and the price policies of OPEC (the Organisation of Petroleum Exporting Countries), it now became a net oil and gas exporter in its own right. Attempts by OPEC to raise oil prices through production controls were now largely successful. Indeed, OPEC officials soon began to realise that when prices rose beyond a certain level, the effect was simply to create the next investment cycle in the shale industry, which in turn would eventually cause another fall in prices.

The most important effect of the shale-gas boom has been, then, a secular fall in oil and particularly natural gas prices.

Natural gas is of particular significance to the renewable energy industry, as it is the main feedstock for gas-fired thermal electricity plants, which compete with wind and solar power. And while oil prices have seen temporary recoveries – with macroeconomics, OPEC production and political and military events still playing a role in influencing prices – natural gas prices have stayed resolutely low due to a global glut and the relative difficulty in transporting gas in its natural state.

When I first started writing on renewables, the issue of competition from cheap gas was a major cause of concern among renewable energy companies and advocates. Instead, the onset of shale gas has helped foster resilience in the industry and drive costs down to grid parity.

Where this can be seen most clearly is the US, the home of shale gas. First, competition from shale gas has created arguably the most cost-efficient wind industry in the world. Helped by excellent wind resources in places like the Texas Panhandle, and straightforward planning and deployment logistics, the US wind industry has constantly driven prices down, using tried and tested turbines with fierce competition from the big manufacturers producing at scale at major US facilities.

Second, big US utilities such as Duke, NRG and MidAmerican have increasingly adopted wind as a core generation technology, often in tandem with producing from gas-fired plants. This is in part due to US rules that stop utilities from hedging prices for their natural gas purchases, which exposes them to price volatility. Wind power's costs are, in contrast, largely predictable across a plant's whole life cycle, providing clear visibility on rates of returns. In other words, the power generation industry in the US takes a 'wind *and* gas' approach, rather than an 'either/or' approach, with wind steadily increasing its share of the overall pie based on its increasing competitiveness. In short, shale has placed a key role in creating a competitive environment for the wind industry, and forcing it to drive down costs in order to present a competitive option for power generators.

Declining government support and the move to auctioning

As we have seen, the wind industry underwent a major boom in the years leading up to the Copenhagen summit. This boom was supported worldwide by different forms of government support for the industry, whether Feed-in-Tariffs, Green Certificates, Renewable Obligations Certificates or other mechanisms. These schemes were extremely successful and supported game-changing investments in technology and the creation of a global supply chain. In Europe in particular, they helped create a number of world-leading companies such as Vestas, Siemens Wind, Gamesa and Enercon.

However they also left the industry exposed to changes in the political arena and particularly in the fiscal climate. The onset of the global financial crisis and the subsequent and long-running Eurozone crisis created a new environment of austerity. Governments began to scrutinise expenditure in energy markets and take urgent measures to reduce or even reverse their commitments.

The most extreme case was Spain, where, as we have seen, support was reined in so abruptly that the industry was left virtually without a market; and Portugal, Romania and other countries took similar measures to create a hard stop in deployment. More widely, governments around the world took measures to reduce levels of support and redesign regulation to foster a more price competitive environment, with a view to eventually ending support altogether. In many countries, this took the form of auction or tender systems to allocate new capacity, and this has resulted in dramatic price falls. Over the last year alone, the amount of auctioned renewable electricity last year was triple that in 2015, according to Bloomberg New Energy Finance (quoted in Clark 2017b).

In Europe, countries moving to competitive auctioning include Germany, Italy, the UK, Spain, as well as Denmark and the Netherlands for offshore wind allocations.

A wide number of emerging market countries also created auctioning systems, including Morocco, South Africa, Brazil, Argentina, Mexico, India and Chile, and as we have seen, these have been spectacularly effective in creating new low price records for wind-power plants.

In some cases, as we have seen, these auctions have brought wind power directly into competition with incumbent fossil-fuel competitors and wind has been able to perform extremely successfully. The auctioning environment creates new challenges and potential pitfalls – in particular, ultra-low prices bid by inexperienced or financially weak companies, combined with unexpected project delays or macroeconomic changes can lead to failed projects or create a climate where companies make opportunistic bids aimed at selling the projects on, soon after tender. However, well-designed tender systems can mitigate against these dangers.

Overall, the move towards competitive auctioning has had a dramatic effect in reducing costs and perhaps even more importantly in creating transparency around the true competitiveness of the wind industry. For years, incumbents and hostile commentators have been able to brand wind as 'expensive' compared to other generation technologies. Now, wind industry advocates are able to simply point policy makers, regulators and commentators to the numbers from the last auction, which provide clear points of comparison with non-renewable alternatives and an indication of which will be the most competitive technologies in the future.

Technology innovation

Behind the rapid falls in costs, is a story of sustained investment in the wind industry and constant technological innovation.

For the wind industry, the path of technology development in recent years has not taken the form of a revolutionary change or breakthrough – the basic configuration of wind turbines remains widely the same as it was two decades ago. However a series of incremental changes has taken place that has hugely increased the efficiency of wind turbines.

The most visible is an increase in rotor size, which allows wind turbines to capture more energy without corresponding increases in the size of gearboxes generators and other components; and this also allows higher capture of more energy from lower wind-speed sites. For example nowadays it is the norm, in the US Midwest market, that projects have capacity factors above 50 per cent, something that was unheard of ten years ago.

Industry founder and former Siemens CTO Henrik Stiesdal points out that behind this move to larger rotors is a set of less-visible factors – the introduction and validation of better calculation tools, which permitted better assurance in the load calculations, and higher-performance blade structures with improved manufacturing methods.

Coupled with the increases in rotor size are huge improvements in gearbox reliability and predictive maintenance, much of this enabled by increasing digitalisation. This latter area is not only allowing better monitoring of turbines, but creating intelligent wind farms that allow turbines to communicate between themselves to better predict and capture wind flow and energy.

Stiesdal predicts that as the industry goes forward and continues to push costs down, some key technology developments will include improved aerodynamics; improved forward-sensing tools to prevent the rotor from bearing the largest operational extreme loads; improved generator topologies for direct-drive machines; new concepts for taller towers; and improved power conversion arrangements. In addition, for the offshore sector specifically, Stiesdal predicts that greater industrialisation of fixed and floating foundations, and 66kV grid connections that eliminate the need for hugely expensive offshore substations will be key.

Industry consolidation, cheaper turbines and components

Technology innovation has been important, but so has the effect of industry competition, the creation of economies of scale and better supply chain management and logistics.

To give an idea of the contrast with the current period, in the boom period around 2008, turbine supply was unable to keep up with demand and customers often faced long waiting periods to obtain turbines, giving companies like Vestas and GE a strong advantage in the market and pushing prices up. This high price environment exerted a strong retarding effect on cost reduction progress. As we have seen, a sellers' market in the late noughties switched to a buyers' market by the early part of the next decade, forcing companies to take existential decisions to reduce overheads and other costs through systematic 'cost out' programmes, sometimes under specialists specifically brought into the company for this purpose.

As we have seen, it also set off a wave of closures, mergers and acquisitions that continues today, and whose implications are only just starting to become clear.

The cost out process was not restricted to the turbine OEMs, but has occurred in parallel within the thousands of companies that supply their components. This intense competition and emphasis on quality control has also played a major role, together with globalisation and the growth of the wind industry in China, which has allowed the incorporation of competitively priced key components from that country across the world.

Finance

Along with cheaper turbines and components has come cheaper finance. Renewable energy projects have long stopped being seen as an exotic or high-risk asset class, resulting in the entry of new types of actor into the sector, including a large amount of banks, private equity funds, and institutional investors such as pension funds. The structure of wind-energy projects, with long life cycles and predictable long-term revenue flows, should make them well suited for such investors, which are now prominent even in the more novel area of offshore wind. Other actors include corporations looking to source their energy needs from wind power (we will look at this group in the next section), insurers and so-called 'impact investors' such as foundations and other socially conscious or philanthropic institutions.And of course, it is important to note that as well as benefiting from a wider universe of investors involved in the wind industry, the industry has been favoured by an unprecedented period of low interest rates globally.

The rise of the corporate PPA

One feature of the new competitive world for wind is the rise of private sale agreements between wind-energy producers and private corporations known as

Corporate PPAs (Power Purchase Agreements) and direct investments by these corporations in wind projects. The roster of corporations signing wind PPAs is diverse, as are their reasons for doing so. The most prominent participants are large-scale technology and consumer goods companies such as Google, Amazon, Microsoft, IKEA, Wallmart, Kirkbi (Lego), Heineken, BT and General Motors.

Power-hungry technology companies and manufacturers, are trying to lock-in their power costs, protect themselves from power price volatility and meet self-imposed sustainability targets. But those getting involved are often doing so for far more prosaic reasons – for example Mexican or Indian small manufacturers or food processing ventures that no longer want to be subject to frequent 'brown-outs' or blackouts, which are a feature of over-stretched electricity systems, or that cannot obtain guaranteed power for new factories or capacity expansion.

An example of the new corporate appetite for renewables is the RE100, a group of companies whose aim is to share best practices and move towards supplying 100 per cent of their power needs from renewables. FTI Intelligence estimates that by the end of 2016, the 31 RE100 companies will have entered into contracts with more than 11GW of renewable energy capacity as they head towards 100 per cent renewable supply by 2030. Looking further on, the 31 companies will enter into contracts with 49.6GW of renewable projects by 2035, equivalent to an average growth of 2GW per annum. And of course, this only represents a small portion of the potential global corporate PPA market by that time.

Within the RE100 group there are 13 technology companies, 42 per cent of the total, but these account for more than 75 per cent of current renewable energy purchases (equivalent to 8.4GW in 2016). Google alone contributes around one-third of this capacity. It is the largest corporate buyer of renewable energy and the sixth largest renewable electricity off-taker in the United States, with 2.6GW directly purchased in 2016. Google is the driving force behind the presence of technology companies in renewable energy purchases, and was aiming to already reach its commitment to reach 100 per cent of consumption from renewables in 2017. With their typically high Corporate Social Responsibility (CSR) commitments, appetite to innovate, but high exposure to energy costs (which can represent one of their key operating cost items especially those with data centres), technology companies are expected to continue to be a key corporate source of demand for renewable energy due to their relatively high cost of electricity purchases as a proportion of total operating expenditure. Google estimates that it consumes 5.7 TWh of electricity each year – equivalent to the city of San Francisco. Managing the amount and price volatility of energy purchases is therefore often high on the agenda for these organisations.

An important factor is branding and reputational risk. As with technology companies, sectors such as consumer goods, retail and financial services are public facing. With growing interest in sustainability issues amongst consumers, many public facing companies see renewable energy as a potential route to demonstrate that they are committed to sustainable operating practices.

However, while technology and consumer goods companies have been the pioneers of the corporate PPA market, the real prize lies with traditional heavy, energy-intensive industry, as they, too, increasingly adopt renewable energy, along with other technologies such as demand-side response and storage to manage their energy costs.

Wind power reaches its "iPhone moment"

Together these developments have decisively changed the cost structure of the wind industry and its position within the global power industry. Of course, huge challenges remain and investment needs to be stepped up still further if wind is to play its full role in achieving the complete decarbonisation of the electricity sector by 2030 as needs to happen if global climate targets are to be met.

Gas and – and far more dangerously – coal, are still seemingly attractive technologies in many emerging and fast-growing economies, depending on the availability and pricing of the fuel, and finance is still widely available. In developed country markets, fossil fuels are not being widely installed, but it is hard to shift assets that potentially still have many more years of life from the grid. Government regulators and grid companies can still be resistant to change due to concerns over disruption to markets and supply stability. And a few backward-looking governments – at present the Trump administration in the US springs to mind – may try and ignore the global scientific consensus on climate change itself or promote spurious arguments over the visual impact of wind.

But in broad terms, the onset of grid parity means that the writing is truly on the wall for the fossil generation, and arguably for the nuclear industry as well. As Mainstream Renewable Power CEO Eddie O'Connor said recently, wind and solar power has reached its tipping point, something he calls its 'iPhone moment'.

O'Connor points out that the first iPhone just over a decade ago used already available technology, and didn't contain a single defendable patent, but the way the package was put together was revolutionary. Within a year of release, Apple had overtaken Nokia in sales and market capitalisation to become the dominant force in mobile phone technology.

Technology advances and accumulated expertise in construction and deployment of wind and solar energy have come together in a way that has created a true tipping point in the energy industry, as renewables costs fall below those of the fossil-fuel family's cheapest option – coal. The effect is that renewables have become the ultimate disruptor in the energy industry, just the way the iPhone did with mobile telephony. Summing up, O'Connor says 'Fossil fuels have lost, the rest of the world just doesn't realise it yet'.

Challenges remain of course, and we shall look at these in the next chapter.

10 Challenges for the wind-turbine industry

As we have seen, the period of 2010–13 was the trigger for a wholesale reset of the wind-turbine industry, which led to the emergence of bigger, more efficient, more resilient companies and helped create the conditions for a renewed, more sustainable period of growth.

However, it would be wrong to imagine that it is going to be plain sailing for the wind industry in the years ahead. In fact, as the implications of the now unstoppable transition to clean energy start to become clear, those wind companies that intend to be successful in the new market will be forced to undergo a far wider process of transformation.

I want to look at some of the key challenges facing the wind sector by identifying some of the unresolved contradictions at work in the industry, as well discuss some new features and developments in the market and political landscape.

Political and market volatility versus globalised manufacturing

The first challenge we have seen through the history of wind energy is market volatility. The growth of the wind industry has been a story of spectacular – followed by steady – growth. But the global figures mask a series of stop–start patterns that can be extremely abrupt. As Alstom's former wind boss Alfonso Faubel puts it when comparing the wind industry to the global automobile manufacturing industry, 'One of the biggest challenges we have is lack of consistent volumes across the board. Automotive has consistent volumes' (Faubel, personal communication).

Growth rates in the more mature markets have slowed, while new countries are continually coming into the market. Given wind's complex interaction with incumbent power-generation complexes, carbon and the energy-subsidy landscape, in the majority of cases the key variable until now has been the prevailing level of political support, which in turn is linked to economic cycles and the level of fiscal solvency of different countries.

Within the EU, the longest-standing wind market, there has been steady growth in onshore wind in Germany and Denmark, where political support has been strongest. Other countries like Spain, Portugal and Greece have seen precipitous growth followed by a virtual standstill in the aftermath of the

Eurozone crisis, with Italy also taking measures to slow down market development. Steady but slower growth has also been the norm in France and the UK, although the former could now be set for a significant acceleration in growth, while the latter is now seeing a politically provoked hiatus in its onshore wind market. Newer markets, such as Romania and Poland in Eastern Europe, have seen extremely fast ramp-ups followed by sharp slowdowns as governments have realised the short-term impact on the public finances, and that the political environment has changed.

Meanwhile, in the US, the creation of the PTC scheme as the preferred tool for supporting the country's renewables sector and the absence of any federal portfolio requirement has led to the most extreme boom–bust cycle in the industry. Changes in the regulatory framework also caused an abrupt fall in growth rates in India, when the accelerated-depreciation and GBI incentive schemes were allowed to expire.

While political volatility is without a doubt the main issue, it is not the only one. In China, the lack of available grid connections, followed by widespread curtailment, were the primary factors in slowing down growth after the spectacular 2008–11 period. The fiasco around delays in building grid connections also virtually halted growth in Germany's offshore sector in 2012, and played a key role in curtailing massive planned investments in Romania. Brazil has been threatened by similar issues, with around 1.2GW of wind farms not connected to the grid at the time of the first edition of this book. On a wider level, a continued lack of interconnections between European countries has created bottlenecks that have prevented the wind industries in some countries from maintaining growth by exporting power.

Another factor behind volatility has been the availability of private finance. The subprime/Eurozone crises in 2008–11 restricted financing for wind projects in the US and Europe. The Chinese government subsequently moved to restrict funding by state-owned banks, also affecting developers. In Brazil, restrictions on the ability of the state-owned BNDES to finance projects had a serious impact on the wind market in 2015–17.

In general, however, finance has become far less of an issue in most markets, with a relatively large amount of money chasing a restricted number of projects. This does not always lead to stable conditions, however, as we can see from the US 'yield-co' model, which seemed to offer a dynamic new model for renewables financing, only to then cause one of the biggest series of company failures at a developer/IPP level in the sector.

Bouts of volatility can cause severe problems for wind-turbine companies. Firms that have invested heavily in new manufacturing plants, sales and service operations to capture fast growth can find themselves scrambling to mothball capacity and relocate or lay off staff a few years later. Think, for example, of Spain's Gamesa, that effectively found itself sitting on factories with nothing to produce from 2011 onwards; Vestas, whose shiny new US factory complex was virtually idle for much of 2012, once the rush of orders that had been supplied that year were delivered; or GE, which at the time of writing the second edition of this book was sitting on three big Brazilian factories facing a dwindling amount of orders.

Wind-turbine companies have taken different approaches to the problem. The most common, and one that has been pursued over most of the last decade by heavyweight OEMs such as Vestas, Siemens, Suzlon and Gamesa, has been to deal with volatility by increasing rather than decreasing its level of globalisation. Former Vestas CEO Ditlev Engel summed up the philosophy when he told me in 2012, 'We need more globalisation and not less' (Engel, personal communication). According to this strategy, being in more markets provides a hedge against constant changes in the rate of growth in different countries. Producing and selling in one market means that if it fails, it is effectively game over. Producing in a wide variety of markets means that you can capture new growth, while having a diverse enough flow of revenues that one stalling market will not kill you. More local production also means less exposure to volatile currency exchange rates, and means that you are more able to influence the local business environment.

Globalisation is also widely considered key to the wind industry's ambition to bring about a sustained reduction in turbine production costs, and therefore in the cost of energy from wind power. Lowering turbine costs requires both a globalised supply chain for components and the kind of scale that only large global manufacturing players can bring about. According to the proponents of this argument – and we will look at this in more detail further on – the creation of the necessary scale requires industry consolidation and the growth of a small number of big players, as the wind industry follows the path of the automobile sector and other major industries.

In practice, however, the globalisation strategy has been hard for companies to follow. This is particularly the case for the 'pure-play' wind-turbine companies that are not linked to wider industrial engineering groups. Restricted balance sheets and the pressure to produce consistently positive EBIT has meant that most companies have found it hard to sustain the level of investments necessary for anything like true globalisation.

The company that has come closest in terms of the number of manufacturing facilities and global sales network is Vestas. To this day, the Danish company is often the best placed to take advantage of newly emerging wind markets, as well as uniquely winning a major share of orders across the three key areas of Europe, the US and Asia. Vestas' capabilities will now be challenged or even surpassed by Siemens-Gamesa (the new global wind-power giant, which we discussed in Chapter 8), while GE has grown its global capabilities through the Alstom and LM Windpower acquisitions.

However, as the most global company, Vestas has also suffered more than most from overcapacity and the cost of mothballing and restarting manufacturing operation. It has also found it particularly hard to cope with the US stop–go cycle. The company invested fully US\$1bn in a complex of manufacturing plants in the state of Colorado in 2010, only to have to mothball them months after completion. After having laid off and compensated a large number of staff, it found itself having to re-hire many of them to deal with the rush of deliveries it faced in 2012 – only to once again have to virtually mothball factories during the PTC hangover lull of 2013. Vestas has had numerous other experiences of setting up production facilities – the Isle of Wight in the UK, Tasmania in Australia, Spain and Italy spring to mind – only to shut them down because of a 'lack of market' a few years later.

Vestas' competitors – Gamesa, Nordex, Acciona and Alstom to name a few – have also found setting up in the US to be difficult, while China has been an even tougher proposition. Suzlon tried hard to compete on terms with its Chinese rivals, with Chairman Tulsi Tanti once famously telling non-Chinese companies setting up in the market to 'stop complaining and start competing' (Backwell 2010b) in China. However, by 2012, both Suzlon and its REpower subsidiary were out. Joint ventures set up by Acciona, GE and Siemens all ended in failure, while Nordex tested the water and decided that selling in the Chinese market was a bridge too far. By 2016, only Vestas, Gamesa, GE – were still attempting to stay the course and treat China as a sales market and not just a place to source components.

Being able to compete everywhere is not a given. India has proved to be an extremely hard place to do business, due to the particularities of the local business environment. On the whole, only those with deep local experience and a willingness to get involved in wind-farm development and land deals have been able to command a sizeable slice of the market. Vestas found that it lost market share quickly when it moved to a supply-only model, and among non-Indian companies only Gamesa – precisely because of its willingness to adapt to local conditions – has been able to significantly expand. This situation is changing once again however with the move towards an auction model, and several manufacturers – including Vestas have re-entered the market.

Brazil may also turn out to be an expensive trap for some, with at least eight wind-turbine manufacturers – including Vestas, GE, Gamesa, Alstom, Enercon, Acciona, Suzlon and local firm WEG – piling in to one of the world's most competitive markets in price terms, and one that is unlikely – at least any time soon – to add up to more than 3GW of installations annually.

Betting on niches

A second option is to pursue a niche strategy, concentrating on one or more key markets where your company feels it has an advantage.

The most obvious example is German manufacturer, Enercon, which has consistently won the lion's share – 39.9 per cent in 2016 according to DEWI– of its home onshore wind market with an expensive but top-quality product and an unparalleled service model. Outside Germany, Enercon has been successful in being an early entrant into some other markets, usually in response to government tendering, including Brazil, Canada and Portugal, but it has often found itself losing out later on as more competitors enter, and it has stayed out of major markets such as the US and China.

While some of its international policies seem to have been due as much to management eccentricities as to strategy, it is hard to argue with Enercon's business model. Selling in Germany, has allowed it to charge premium prices; and Germany's onshore market has been the most stable, steadily growing wind market in Europe, and indeed the world. Enercon has not faced the brushes with bankruptcy that competitors such as Vestas or Gamesa have faced. Its technology and products are widely admired and popular among its customers.

On the other hand, Enercon is exposed. During the recent political debates around Germany's *Energiewende*, it was interesting to note both the sudden willingness of traditionally shy Enercon company officials to engage with the media and policy debates – and their nervousness. As one competitor pointed out, 'They have more to lose from things going wrong here than any of us' (personal communication).

Other companies following a niche strategy are Nordex, which under the leadership of CEO Jurgen Zeschky pulled out of competition in China and the US, and concentrated in building market share in its home market of Germany – while spotting niche opportunities in places like Turkey, Pakistan, Sweden and Uruguay. Referring to the company's downsizing of its international manufacturing footprint, Zeschky said in his letter to shareholders in the company's 2012 report:

> I am confident that with the leaner structures which we have implemented we will be able to make full use of the advantages which we have in the market place. This is because we have always been successful as a mid-size company. Customers do not expect Nordex to be a big corporation but, rather, a flexible engineering partner that is able to respond quickly and understands all aspects of their business.
>
> (Nordex 2012: 5)

Today's market environment has caused analysts to question whether the specialised, niche players will be able to continue to prosper. As we have seen, the Nordex shareholders eventually decided to 'go big' and opt for a takeover of Acciona Windpower with a view to targeting a place in the Top Five.

Analyst Eddie Rae predicts that Enercon could end up being squeezed in its home market in Germany as giant competitors like GE target the market. The transition in Germany from markets based on Feed-in-Tariffs to an auctioning structure means that price will become as important as quality and that 'buy German' will be less important, exposing the company to harsher competition. Others are more optimistic about the perspectives for medium-sized players. Industry founder and former Siemens Wind CTO Henrik Stiesdal says he is sure that there will be a place for Enercon and other mid-size players, 'provided they focus their strategies and offerings'.

In offshore as well, the market does not seem to be favouring the existence of specialised niche players. As we have seen, the French-owned nuclear giant fought an unsuccessful battle with India's Suzlon to take over German turbine company REpower in 2007. It then took over the smaller German designer and manufacturer Multibrid and built up a manufacturing and marketing operation focused solely on the offshore space. A few years later, the offshore market is dominated by a few global companies – Siemens, MHI-Vestas, GE – active in their own right or through their owners in onshore wind and the wider energy sector.

Of course the divide between 'fully globalised' and 'niche players' is far from clear-cut. As an example, Tier 1 player GE has pursued a highly selective approach to the markets it enters, while making sure it maintains leadership in its home market of the US, helped as we shall see by synergies within the GE group. GE has an extremely lean manufacturing footprint that relies on complex logistics to deliver projects. The main components for the giant Fantanele wind farm in Romania, for example, came from around the world; blades from Brazil, nacelles from Germany; towers segments from China; and many internal electrical components from the United States.

Spanish-owned turbine manufacturer Acciona Windpower largely scaled back its production to its plants in Spain, avoiding overreach and keeping its core facilities viable. However, being a niche manufacturer didn't stop Acciona Windpower from starting up new facilities to supply third party customers where it sees a market opportunity, for instance making major investments to meet stringent local content requirements in Brazil.

And it is worth looking at Germany's Siemens, which has historically derived a large part of its total wind-turbine sales from offshore, even as it pursues a strategy of trying to increase its market share in the main onshore wind-turbine markets.

Overcapacity and consolidation

The expansion of global manufacturing operations by OEMs and the entry of new manufacturers into the market led to a situation of over 100 per cent overcapacity in the market by 2012, which inevitably led to lower turbine prices and a prolonged squeeze on turbine manufacturers' margins.

According to Robert Clover, who headed research into wind-sector companies for investment bank HSBC and then went on to join FTI Consulting, there was around 85GW of manufacturing capacity for turbine nacelles compared to a demand of around 40GW in 2011–12. He notes that a significant part of the excess capacity built during the period was in China, where, as we have seen, there was a large number of OEMs in the market.

Clover identifies a number of factors behind the build-up of excess capacity in 2008–10. The main causes were the high returns in the wind industry, which attracted large amounts of capital, and the simultaneous onset of an economic slowdown, slower growth in power demand and regulatory pushback. 'The wind industry before the crash expected compound annual growth rates of something like 20–25% over the 2008–2013 period', he says, 'and this ended up being more like 6.5%' (Clover, personal communication). He adds that 'The excess capacity really built up in 2008–2010 as investments planned before the crash came online.' Indeed 2008–10 saw new companies continuing to enter the market and build new plants despite the large number of already established players in the market.

'We went very fast from a situation of excess demand to excess supply; from high prices to low prices; from long delivery times to manufacturers chasing customers – even those smaller ones that they had dismissed in the race to scale in the upturn', says former EWEA CEO Christian Kjaer (personal communication).

Fast forward a few years to the record year of 2015 and demand was 63GW. There have been substantial reductions in global turbine manufacturing in places like Spain since 2012, but overcapacity is still estimated at around 40–50 per cent.

As we have seen, when the first edition of this book was published, something like a 'phoney war' was occurring, with analysts and industry executives all stating that consolidation was necessary, but nobody willing to make the first move, as companies did not want to acquire more under-utilised factories in a market with 100 per cent overcapacity.

When I asked GE's Vice-President for Renewables, Vic Abate, in 2012 why GE didn't simply buy out its cash-strapped competitors, he replied by asking why GE should buy excess capacity in a situation of 'rampant overcapacity', adding there was a danger of simply acquiring out-of-date industrial technology. 'When you think of consolidation you think of a buyer and a seller', he says, 'but I ask myself "what can't I do organically?"' (Backwell 2013f).

Alfonso Faubel, then Alstom's wind power boss, asked whether companies would be able to simply swallow up their smaller competitors, said, 'You need to preserve shareholder value. Meaning you can't just spread around the world with new factories; it doesn't work. You need to use cash very carefully' (Faubel, personal communication).

Ironically both companies would merge a few years later, when the big wave of wind sector M&A finally emerged.

As I predicted in the first edition of this book, economic recovery, the availability of cheap money struggling to find a competitive return and an increase in turbine demand was 'likely to make consolidation more, and not less likely to occur, as increasing allow[s] the wind divisions of the major industrial players to persuade cautious head offices that a new wave of investment in wind is a good bet', and this is effectively what has happened.

Alongside the process of consolidation among OEMs, has been a similar process among key component suppliers, with a combination of large-scale acquisitions – for example in gearboxes – and bankruptcies. FTI Consulting estimates that wind industry saw 129 suppliers collapse or leave in the 2013–15, of which 88 were from Asia, 23 from Europe, and 18 from North America (FTI Intelligence 2015).

Supply chain expert Eddie Rae, says that consolidation of the supply chain for critical components such as gearboxes is a natural process as the OEMs look to work with supply partners who have the mass, financial strength, global footprint, process sophistication and product innovation required to 'follow' the OEMs in their own growth development. He predicts that in the future there will probably be only three to five gearbox players and the same number of blade manufacturers, although he says that in the latter area much will depend on the ramifications of GE's takeover of LM Windpower, which could even spur the emergence of a new player in the market.

The question of what level of spare capacity the industry needs and can afford to maintain, and how many companies will survive, is still very much pending. It is impossible to predict how many turbine manufacturers are likely to be in the market by the end of the decade. Some analysts predict that there will be around ten to fifteen significant companies, compared to the dozens of players operating today. And within this, there is likely to be a leading group of three truly global companies made up of Siemens-Gamesa, Vestas and GE. China's market will continue to consolidate around a small group of big players led by Goldwind and Envision, but given that China's wind market will see slower growth for the next few years, it is unlikely that they will be in a position to challenge for the very top global spots again until the early 2020s.

For industry founder Henrik Stiesdal, the wave of mergers that took place in 2015–17 were an example of 'classical consolidation in a maturing industry … Similar bursts of consolidation are known from many other maturing industries', Stiesdal points out. 'The automotive industry of the 1910s and 1920s is a well-known example, but lesser-known examples may be more similar to the wind industry, for example the consolidations within the steam- and gas-turbine industries.' However, Stiesdal cautions that the mergers are not of a magnitude that will 'substantially alter the levels of industrialisation. … The merged companies are likely to maintain individual factories more or less unchanged', says Stiesdal. 'However, the merged companies will have more muscles and more volume to push forward industrialisation in the coming years, and I am sure that five years from now we will see clear examples of even more mature supply systems." (Stiesdal, personal communication, May 2017).

Vertical integration versus the networked industry

Related to market growth and industry capacity is the issue of what kind of business model wind-turbine companies should follow.

For years, the model for the big Tier-1 wind-turbine companies was 'vertical integration', a management system that emphasises control through different levels of the supply chain. Vertical integration is used, for example in the oil industry, to describe companies that combine both 'upstream' oil exploration and 'downstream' oil refining and marketing.

In the wind industry, the major wind-turbine companies have controlled manufacturing of everything from steel hubs through to gearboxes, electrical components such as generators and control systems through to towers, as well as the blades and nacelles which most industry officials agree make up the core of the business.

Tier 1

Vestas.

SIEMENS Gamesa
RENEWABLE ENERGY

GE

Tier 2

GOLDWIND

NORDEX — acciona

ENERCON
ENERGY FOR THE WORLD

ENVISION

Tier 3

MINGYANG ELECTRIC
明阳电气

上海电气
SHANGHAI ELECTRIC

CSIC

XEMC
湘电集团

DEC 东方电气
DONGFANG ELECTRIC

联合动力
United Power

SENVION   SUZLON

Figure 10.1 Possible OEM market landscape in 2020 (Source: FTI Consulting).

Vertical integration became the norm during the boom years of the wind industry in 2005–09, when turbine manufacturers were fighting to ensure that components were available to them during periods of high demand. Leading manufacturers carried out a series of acquisitions of supply-chain companies or acquired the necessary technology to make the components in-house. In 2013, Jukka-Pekka Makinen, CEO of leading power electronic supplier The Switch (sold to Japanese industrial group Yaskawa in 2014) said: 'Today there is a clear reversal of that trend. ... Turbine companies are concentrating on their core competencies, while outsourcing production, because they have come to understand that this is the only way to increase supply-chain flexibility.' (Makinen, personal communication).

Makinen advocates a 'networked' approach, with open collaboration between companies being the new industry norm. 'What's new in the networked business environment is that all parties involved are sitting around the same table, working towards the same result', he says (ibid.), adding that the key to successful collaboration is building trust among all parties.

Makinen's appeal for openness is far from being the accepted norm among the big turbine producers, some of whom are highly secretive about their technologies. Tensions still arise continuously between turbine OEMs and their component suppliers – the most common cause of high-profile turbine failure continues to be non-performing bought-in components – as both sides attempt to maximise returns, and this is likely to be the case for the foreseeable future.

It is clear, however, that under conditions of spare capacity companies reassessed what they saw as core business, while looking to squeeze more value from their operations by shedding operations where they do not enjoy any specialised competencies.

Leading turbine manufacturer Vestas carried out an extensive outsourcing programme, including selling off a major tower factory in Denmark and its metal-casting manufacturing, and outsourcing logistics, warehousing and transport operations. The company also outsourced key components for its flagship V164 8MW turbine. It also began to use its tower production facility in Colorado, USA to produce towers for third parties, to minimise spare capacity and bring in extra revenues.

There is of course no clear-cut divide between companies taking a vertically integrated approach and those taking a more flexible approach to the supply chain, and in a buoyant market, manufacturers will be tempted to extend direct control and become more, not less, vertically integrated. US turbine producer GE, for example, for a long time brought its blades from Brazilian supplier Tecsis through long-term supply deals, and others have a mix of in-house and third-party-produced blades. Now, however, GE has made a major move into blade production, acquiring the world's biggest non-OEM manufacturer, LM Windpower in 2017, and modular blades specialist Blade Dynamics in 2015.

The debate about outsourcing is part of a wider discussion about what model the wind industry should follow for industrialisation. As we have noted, auto manufacturing has been seen as a key model by company executives, and over the last couple of years, companies have hired a number of managers from the auto industry.

In June 2012, Vestas appointed Jean-Marc Lechêne – former director of Michelin's heavyweight tyres unit in Europe – as its chief operating officer, with responsibility for manufacturing and sourcing. At the same time, Spain's Gamesa appointed Xabier Etxeberria to the new role of 'business chief executive officer'. Etxeberria comes from the automotive arm of engineering group GKN, where he spent ten years in the Driveline division. Etxeberria was employed by Gamesa's Executive Chairman Ignacio Martín, who himself worked at GKN and more recently was Executive Vice-Chairman of car industry component and sub-assembly supplier CIE Automotive. French industrial giant Alstom appointed Alfonso Faubel – who had held senior positions in US-owned auto component and technology supplier Delphi – as head of its wind business back in 2009.

It is worth pointing out that the level of mechanisation within the sector remains relatively low. Most nacelle assembly plants involve a large warehouse with groups of workers moving around from workstation to workstation fixing the different sets of components into place, before the finished product is hauled out and loaded onto its transport.

Alfonso Faubel, a strong advocate for modernising production systems, describes the industry in transition from being a 'Mom-and-Pop shop kind of industry, to a truly industrial set up where automation, where building quality, where all best practices from supply-chain management were being utilised' (Faubel, personal communication).

An interesting case is that of Siemens, which, like Alstom, has extensive company-wide experience it can bring to bear from large-scale industrial manufacturing. Siemens was the first wind-turbine manufacturer to introduce a true moving production line in Brande in 2010, using a flow production concept designed by Dürr Consulting. Siemens said the system reduced assembly time per nacelle to 19 hours from 36 hours, and the number of workstations from 18 to 8. Dürr said the digital three-dimensional factory layout it completed in April 2009 'forms the essential basis for all new Siemens wind-turbine assembly lines all over the world, especially in the Asian and American regions' (Backwell 2010c) and allows Siemens to globalise its nacelle-manufacturing operations and set up factories wherever there is demand.

Former Siemens Wind CTO Henrik Stiesdal said that he looks closely at the truck manufacturing business, which he says provides a more realistic point of comparison with wind than the consumer automobile industry in terms of the number of units produced. He predicts a growing level of automation in production, although he says that the industry's likely maximum scale implies that full robotisation is not on the cards.

Siemens' strategy is based on the automotive industry's practice of mass-produced modular components, while wind turbines are 'bundled' into product platforms. 'We can reduce production and logistics cost by standardising and modularising components within our product platforms', said Stiesdal, describing the strategy as 'a major step towards achieving our goal of making wind power independent from subsidies' (Stiesdal, personal communication).

For Stiesdal, the main barrier to automation in nacelle manufacturing is the amount of work involved per unit. 'The possible benefit from automation is much more likely to be at the level of the component suppliers', he says. 'For blades it's different and this is where we are focusing the automation effort' (ibid.).

True to his profile as one of the wind industry's most consistently independent figures, Stiesdal says that what companies actually end up physically producing will continue to be debated, and he even puts forward the idea that a company like Siemens could still be a turbine manufacturer without actually producing nacelles. 'You can actually imagine companies being more "knowledge based" rather than owning industrial assets', he says. However, he adds 'there is no obvious best strategy – Enercon is equally successful with a very high degree of vertical integration' (ibid.).

While the general thrust of big industry players is towards automation, we have seen that there are powerful forces that have put limits on the intensity of capital and processes like automation, the most important being market volatility, which makes manufacturers reluctant to make big new investments in plant.

Another factor is the common use of local content requirements, as national government's try to 'capture' wind-power investment and employment. Rising markets such as Brazil, Turkey and Canada all have some kind of local content requirements, as have previous growth stars such as Spain and Portugal – whether these are through government tenders, or through restrictions on financing as is the case in Brazil. In the most basic terms, the demand for local manufacturing – as socially and politically necessary as it may be – can prevent components for wind turbines being produced at the lowest possible cost and mitigate companies' attempts to introduce real scale advantages due to the limited size of individual markets.

Wind-turbine companies, which were obliged to set up local manufacturing to address the fast-growing Brazilian market due to restrictions on BNDES financing for developers (see Chapter 4), aimed to set up local manufacturing in the cheapest and most cost-effective way. However, doubts over the total size of the market in the future meant that they were unlikely to invest in state-of-the-art mechanised plants.

An example of the effects that this environment can have is Vestas' decision in 2012 to change from a system of blade manufacture it had introduced for its V112 turbine in Europe – whose purpose had been mass-standardised production – to a new design based on the tried and tested method of individually crafted blades. The reason for this was the need for a 'capex light' industrial footprint that allows the company to quickly – and cheaply – set up operations in new markets.

Clearly, Vestas' decision made sense for the company, in the context of its offensive to reduce capital expenditure costs. From an industry-wide perspective, however, a promising move towards industrial scale that should have eventually led to a sustained reduction in cost was halted in favour of an earlier technique that is more costly in terms of man-hours per unit.

Again, the tension between global manufacturing scale and local production is not an out-and-out dichotomy. The Switch, for example, championed a system of 'model factories' that can be set up cheaply and quickly in any market, and which it argues also embody best manufacturing practices.

There are clearly still many issues to be worked out over the coming period. One can hope that the consolidation currently underway and an upsurge in market growth will allow companies to give some answers to these questions. Attempts to learn from the automotive industry have, until now, mainly been about how to apply cost-cutting. Getting to the level of sophistication of auto manufacturing, however, is about achieving a higher level of investment over a sustained period and achieving productivity through automation and not just about being leaner.

Diversified industrial or pure-play?

We should consider at this point that dealing with geographical and generalised volatility – i.e., the varying rates of growth across the global sector – in the wind industry is to a great degree a different proposition for the 'non-specialist' wind-turbine companies, that form part of much larger industrial companies. The main factor is the size of company balance sheets. In simple terms, wind-turbine manufacturing losses are not likely to bankrupt parent companies. A loss that could cripple Vestas (revenues €10.2bn) is likely to be less significant for Siemens (revenues €80bn) or GE (US$120bn) (all 2016 numbers).

At the same time, many of the big industrials possess a natural 'hedge' because of their other activities in the wider energy sector. If a company like GE, Siemens or Mitsubishi has a bad year for its wind-turbine equipment division, it may well have had a good year for its division that produces turbines for gas or coal plants.

Being diversified also allows companies to shift personnel to high-activity areas and make use of synergies across manufacturing assets. Famously, GE was able to ride the rollercoaster of US turbine deliveries during the record year of 2012 by assembling wind turbines in its gas-turbine manufacturing plants, allowing it to take first place globally. It was then able to shift personnel back to their usual duties a year later when the wind market entered a lull. Others had no such advantages on either the upward or downward part of the cycle, and faced expensive ramp-ups and then layoffs of personnel.

It is worth remembering, however, that although they have bigger balance sheets, parent companies have shown that they will try to impose strict financial discipline on their wind-turbine divisions to avoid any negative impact of company-wide EBIT targets. This means, even though bigger balance sheets are in play, wind-turbine divisions in practice arguably face similar restrictions on their activities as the specialists. On the other hand, the existence of the parent company means that clients are unlikely to be scared off by doubts about the overall solvency of the turbine supplier, in contrast to what has happened on occasions over the 2011–13 period with some of the 'specialists'.

Eddie Rae summarises the challenge of combining a lean manufacturing base with a diversified presence in the following 'bucket list' for successful OEMs:

- A manufacturing footprint on three continents

- Rapid deployed manufacturing and supply chain flexibility
- Key supplier strategic 'partnership' arrangements
- Smart modular product design, driving down cost and development/process times
- Preferably part of a wider multinational concern
- A diversified in-house and out-sourced supply chain
- Onshore and offshore capability
- 4~5 per cent per year R&D investment
- New technologies savvy (and US 'West Coast' partnering)
- Close project developer links
- O&M task outsourcing
- O&M as a driver of growth
- Strong local market business development and influencing skills/network

Beyond the turbine – taking a 'whole system' approach

But for the future, the challenges are greater than 'simply' producing and selling world-beating wind-turbine equipment. As the world's economy undergoes a wave of technology disruption, and wind and other renewable energy sources supply an ever greater share of power, the winners are likely to be those that can grasp the possibilities created by these changes, and take a 'whole system' approach to the energy business.

'Being a pure-play equipment manufacturer is the world of yesterday, being a sustainable systems solution provider is the world of tomorrow, those that don't adapt to this new market dynamic are placing their business at risk', says Aris Karcanias, a leading wind-sector expert and co-lead of FTI Consulting's Clean Energy Practice.

For Karcanias, changes in the structure of the energy market have profound implications for wind-power companies. The market has moved from on from the cost and financing of front-end capital expenditure, to a common understanding based on the levelised cost of energy across project lifetimes. Customer landscapes are being transformed, as traditional power consumers incorporate the new tool set that new technology is creating, manage their demand in new ways and bypass traditional intermediaries such as distributors.

Meanwhile growing variable generation is forcing governments and energy users to look at 'system LCOE' – i.e., the impact of technology choice on system integration, and move away from measures such as preferential dispatch for renewables and Feed-in-Tariffs, which have favoured the growth of the wind sector over the last three decades, ushering in a new 'world without preferential treatment for variable generation'.

'The single greatest challenge for wind energy and indeed for solar is the variability in output', says Eddie O'Connor, the CEO and Founder of Mainstream Renewable Power (personal communication). He points out that in places like Chile – where, as we have seen, Mainstream is in the process of become one of the country's biggest power suppliers – renewables companies are increasingly having to take a 'whole system' approach and take into the account their impact on the grid.

> The old 'take and pay' model – where wind-power plant was paid for every unit generated no matter if it was needed – can be replaced by one where renewables generators supply 'firm' power on demand. … Rapid advances in storage and demand-side management, as well as forecasting and generation technology, are enabling generators to shoulder the responsibility for providing 'firm' power even in existing systems built for always-on coal power.
> (O'Connor, personal communication)

Vestas CEO Anders Runevad says: 'Looking ahead, the success of energy markets with high penetrations of wind will require technologies working together in energy-system solutions that can store electricity, manage demand, and give stability to the grid in order to increase power output, lower the cost of energy, and address variable generation.' Runevad points out that although 'wind remains our unquestioned core' Vestas is today involved, pursuing and potentially expanding upon initiatives that involves grid integration, hybrid systems, and storage solutions.

For FTI's Karcanias, the wind companies that will survive in the new landscape will be those that are able to ride market imperfections and create new business models and utilise technology to access capacity and ancillary revenues created to maintain security and stability including those to make viable uneconomic yet dispatchable fossil plants. The future is wind co-existing with complementary sources like solar PV and storage – meeting the evolving customer needs as 'intelligent and sustainable system solutions providers'.

'The question in this rapid evolution becomes, "how do I engage with my end customers while working around an aged grid and an imperfect market",' says Karcanias, adding that the old route to market of selling equipment and indeed electricity through utilities alone will be 'obsolete' (Karcanias, personal communication).

Related to this, the emphasis for equipment manufacturers will necessarily shift into a world of OPEX (operating expenditure) as the business becomes more mature and markets and sites become more saturated. Those companies that can offer new financing models such as leasing and make project-level commitments on costs through – for example in power-supply agreements with industrial power users – will be likely to prosper.

'Players that can readily address this underlying end customer need, accept and adapt to market dislocations and are willing to advance disruptive technology into advanced business models that bypass the incumbent aged centralised generation model will be the ones that survive', predicts Karcanias. Some of the major wind-turbine manufacturers have made bold statements and considerable investments in digitalisation, but the race is still on to turn these investments into mainstream revenue-generating products that can solve a real customer need, Karcanias notes.

The new landscape and evolving thinking among wind industry strategists is also likely to affect future consolidation and M&A activity. There is no doubt the wind industry needs and shall continue to innovate its existing portfolio to drive down the LCOE and fund the new journey. However, the route to success could likely lie in the building or acquiring of innovative companies such as power aggregators (or so-called 'virtual power companies') or storage system solution providers, rather than driving a within-industry consolidation. In that vein, we are likely to see a shake-out through exits rather than acquisitions.

As Karcanias, points out, the new world of information has yet to take full effect, indeed those that are open to new ideas and are willing to experiment will be well placed in the coming years to capitalise on this transformation and 'prepare for the digital energy era – a world of automation and artificial intelligence'.

China versus the West – a market split in two?

Another issue that will have a lasting impact on wind-industry productivity and costs is the impact of the Chinese wind industry on the global market. The most striking feature of the current situation is that from one perspective, there have been two distinct and separate wind markets: the Americas and Europe on the one hand and China on the other.

The European and American markets are dominated by the same group of international turbine manufacturers, led by Siemens-Gamesa, Vestas, GE, Enercon, and Nordex-Acciona. The Western companies also share many of the same wind developers/utilities led by Iberdrola, ENEL, EDPR, EDF-EN, E.ON, Mid-American, Vattenfall, Pattern, Invenergy and so on, although there are also many regionally focused ones too. The Chinese market is almost completely dominated by Chinese turbine manufacturers, as we saw in Chapter 3, with the non-Chinese companies struggling to make it into the top ten spots. China's top ten wind developers are entirely Chinese, and none of the biggest international players have even attempted to break into the market.

The existence of two almost separate markets, particularly when China is consistently outstripping any other country or even region in terms of annual growth, has contributed to the growth of unsustainable levels of excess capacity. And it can only hold back achieving higher productivity and lower costs in the industry as a whole. Of course, the issue has more nuances than at first appear. Western companies that have lost market share or even given up on selling into the Chinese market at all often maintain large-scale manufacturing operations in the country and rely on China as the source of many of the critical components that they use around the world.

According to supply chain expert Eddie Rae, a typical OEM supply chain will include roughly 20–25 per cent of turbine value (excluding blades and towers, which are typically purchased in the region where wind farms are commissioned) from China. This is usually in the simpler 'transformational technology' groups like castings, forgings and structural components, but increasingly also from higher technology groupings like drive train and electrical systems as the Chinese manufacturing quality and technology expertise improves.

Companies like Nordex, which decided to pull back from the Chinese market after negotiations over a joint venture with a leading developer failed to prosper in 2012, continue to use China as a major source for component supply, as do others that have pulled back from the market.

GE also maintains significant production activities in China, as well as maintaining a small share of the local market. An official from GE commented recently in Beijing that 'you could be receiving GE turbines made in China and you wouldn't even know it' (personal communication).

The divide between Chinese and non-Chinese OEMs is often not as wide as it may first appear, and questions of intellectual property rights are seldom as conflictive as they have been in the Sinovel–AMSC case. Goldwind, the Chinese company that is arguably best placed to expand internationally, bought into the highest level of turbine design through its acquisition of German designer and manufacturer Vensys, and licensees around the world are using designs whose patents are controlled by the company. Goldwind has also been the most successful of the Chinese companies in gaining experience in hard-to-enter 'mature' non-Chinese wind markets, and to all extents and purposes appears as a Western company in places like the US and Australia.

The push by Chinese turbine manufacturers to become international players will eventually materialise at some level, as we have discussed in Chapter 3. Chinese companies such as Longyuan or Envision are slowly accumulating projects abroad and will translate into a growing foothold for Chinese turbines in international markets. Other important market participants such as State Grid are also stepping up their international presence. And the steady shift of financial power from the West to China – helped by developments such as the soon-to-be establishment of the renminbi as an international currency, and the availability of finance – will continue to increase Chinese companies' competitive advantage in the years to come.

Meanwhile, the Western companies that have decided to stay and fight it out in China could eventually win back some market share, as developers focus on quality, and the costs of local producers rise, although opinion is divided on this, with 'don't hold your breath' the prevailing view from most analysts. One possibility is that a Western company such as Vestas, with its balance sheet now strong again, could reverse the prevailing global trend and acquire a local manufacturer with strong utility links in order to increase its market share in China.

At present though, the divide between the Chinese and non-Chinese wind markets poses a number of problems for the wind industry. No major OEMs have been able to reap the potential scale advantage of rapid growth in *all* the major markets, and neither have we seen true global competition that could really raise productivity and lower costs. Instead, as we have seen, the Chinese market has until now spawned an artificially high number of new entrants, with many of these winning share despite low levels of technical expertise and quality. Non-

Chinese companies have had to scale down their presence in China, while Chinese players have, on the whole, struggled to gain credibility for their products abroad. This means that non-Chinese wind markets have not seen the potential cost benefits that Chinese industrial productivity and scale has brought to, for example, the solar PV module market. Meanwhile China's market has yet to benefit from the experience and skills that developers who are used to making projects viable in highly competitive markets could bring.

Hopefully, the not too distant future will see a truly international turbine market, with highly capitalised and internationally diversified players that are able to compete for leadership in China and internationally.

Solar PV, ally or rival?

For much of the period covered in this book, wind's renewable energy technology 'cousin', Solar PV, has been considered a higher cost energy source that would not be ready to reach grid parity and massive deployment across the globe for some time to come.

Consequently the wind industry has viewed Solar in a relatively benign fashion, rubbing shoulders with the solar industry in renewables industry bodies and often jointly lobbying. The two industries share a common mission in eliminating carbon emissions from global power generation, and developers and IPPs have often been involved across both technologies. Relations have not always been smooth however. One notable example of how things can go bad is Spain, where the wind industry in effect moved to distance itself from the solar sector when the country's government began to attack 'expensive' investment in renewables and seek to revise existing Feed-in-Tariff commitments in 2010.

In 2016, however, the picture looks very different, as the solar industry installed more power capacity than wind for the first time ever with 71GW for solar compared to 51GW for wind according to the International Renewable Energy Agency (IRENA).

According to Bloomberg New Energy Finance, unsubsidised solar is out-competing coal and gas in a growing number of markets, but solar projects are also costing less to build than wind projects in fast-growing emerging markets, although wind remains cheaper in most OECD markets due largely to the relative availability of sunshine and wind (see Randall 2016).

'Solar investment has gone from nothing – literally nothing – like five years ago to quite a lot', says Ethan Zindler, head of US policy analysis at BNEF. 'A huge part of this story is China, which has been rapidly deploying solar' and helping other countries finance their own projects (Randall 2016).

There have been a series of rapid gains for solar PV in price terms in 2016, including a deal for US$29.10/MWh in Chile, which came in significantly lower than bids from the wind sector and prices of US$24.20/MWh in the United Arab Emirates. In May 2017 companies led by SoftBank Group of Japan and Taiwan's Foxconn won contracts in two auctions in India for as little as 3.8 cents a kWh (US$38/MWh), sharply below the previous bids around 5 cents and within striking distance of the lowest so-far prices in Chile and the United Arab Emirates, according to Bloomberg New Energy Finance. Overall, solar prices are down 62 per cent compared to 2009, and a series of technologies being rapidly deployed around the world – from batteries to micro-grids to electric vehicles – will make solar easier to deploy and integrate by mitigating solar PV's major disadvantage of extreme variability. IRENA anticipates a further drop of 43 per cent to 65 per cent for solar costs by 2025. That would bring to 84 per cent the cumulative decline since 2009.

The good news is that the world is likely to see solar becoming more economical than coal in most markets by 2025, and even in China by 2030, and this will increasingly influence the decisions and forward planning of corporations and governments alike. For the wind industry, however, solar PV is emerging as a competitor – and an aggressive one at that – in markets around the world, and this is obliging companies and industry advocates to rethink their public affairs and lobbying strategies.

On the developer/IPP side, cheap solar is simply encouraging companies to diversify their portfolios and enter or re-enter the solar market. Large-scale developers such as ENEL and Mainstream Renewable Power are equally comfortable developing wind or solar and there are big gains from developing sites 'holistically', i.e., taking advantage of grid connections and other infrastructure to install an optimal combination of the two technologies.

However in some markets, wind and solar are competing directly for the same available capacity, and often head-to-head in long-term power tenders. This is likely to cause a major rethink of strategy among wind OEMs over the coming years.

The Trump Effect – Renewed political challenges

As we have seen, political volatility has been one of the major challenges facing the wind industry since the beginning. The historic Paris Agreement at COP21 at the end of 2015 seemed to herald in a new era of international consensus about taking action to combat dangerous climate change, something that was strengthened by rapidly increasing commitment and engagement from the biggest developing economies of China and India.

In the early morning hours of 9 November 2016, a new era of uncertainty emerged seemingly from nowhere – with the election of Donald Trump in the US.

During the course of his 2016 campaign, Trump presented an 'America First' energy platform via clear support for the coal, oil and gas sectors. Future energy challenges, he suggested, would be met by 'opening federal lands for oil and gas production, opening offshore areas, and revoking policies that are imposing restrictions on new exploration technologies'. Combined with prior statements questioning the veracity and impact of climate change, along with the posturing of senior advisors on federal involvement in clean energy, this created an anxious debate within the US renewables industry around whether or not existing federal support mechanisms may be altered – or eliminated altogether.

On 1 June, Trump announced that, as feared, the US was withdrawing from the Paris Agreement. The move caused almost universal condemnation and the US came under heavy pressure from its G20 partners to reverse the move at the organisation's summit in July. Domestically, the move was condemned by major US corporations, as well as by powerful states like California, which promised to forge closer cooperation with China to combat climate change and increase use of renewable energy.

At the time of writing of this second edition it was not clear what the real consequences of the move would be, with media reports suggesting that Trump was seeking to have the US re-join the accord, possibly after some face-saving changes in the terms of its adherence to the Agreement (see Sampathkumar 2017).

Trump's policy bias is not completely unexpected, given the context of GOP ('Grand Old Party', i.e., Republican) rhetoric against climate change and other environmental measures. For example in 2013, the Republican-majority House proposed cutting Environmental Protection Agency (EPA) funding by one-third and, in 2015, put forward 61 anti-environmental bills (including one to cut funding for renewables research by 50 per cent). The Republican Senate has been equally as hostile to renewables – a bill in 2015 aimed at cutting EPA's budget by 50 per cent.

The energy appointments made by President Trump provided a clear indication of the administration's bias towards fossil fuels. Fossil-energy lobbyists Mike McKenna and David Bernhardt were appointed to lead the Department of Energy and Interior transitions, respectively. Vocal climate-change sceptic Myron Ebell led the EPA transition, and Scott Pruitt – a plaintiff in a lawsuit against the EPA – will now lead it. Pruitt's LinkedIn page shows that he identifies himself as 'a leading advocate against the EPA's activist agenda', having previously sued the EPA over fossil-fuel emission regulations.

Texas Governor Rick Perry was appointed Energy Secretary. A highly visible champion of fossil fuels and, in particular, oil, Perry has also expressed scepticism of the science underlying climate change. At the same time, however, Perry, has a prior history of backing renewable energy initiatives in his own state, and in particular, wind power.

Indeed, it is important to note that the majority of states responsible for producing most of America's renewable energy have Republican senators.

'Regardless of party loyalties, it is difficult for us to imagine any politician, regardless of their political stripes, supporting wholesale changes to environmental legislation that might have significant negative impact on job creation, job retention, and tax revenue in their constituent states', says FTI Consulting's Robbie Goffin in a research note following Trump's election (Goffin *et al.* 2016).

It is worth noting also that 80 per cent of all wind farms reside in Trump-voting states (such as Oklahoma, Kansas, Nebraska and the Dakotas), and 77 per cent of Trump supporters want to see more wind farms in the United States. Wind power has solid bipartisan support as a significant employer (88,000 jobs alone in the Rust Belt). Thus, while traditional GOP posturing on renewables has been broadly unsupportive, the reality is that measures that could significantly damage the renewables industry are likely to be diluted by Republican lawmakers themselves.

Major US corporations such as GE will no doubt be using their weight to lobby for continued support for the wind industry, as will industrial companies such as Tesla, technology companies such as Google, Amazon, and big utility investors such as Warren Buffet.

At the time of writing, the outlook for the US wind market outlook in the near term appears to be stable, as President Trump's the new Treasury Secretary Stephen Mnuchin has confirmed support for the smooth phase-out of the existing Production Tax Credit ('PTC').

However, US market development in the medium term remains uncertain after Trump issued an executive decree instructing the EPA to review the legislation that created the Clean Power Plan – already mired in a legal battle – and decide whether to 'suspend, rescind, or revise it'. While 17 US states mounted a legal challenge to the executive order, this meant that effectively the CPP, which created a stable framework development of renewable energy in the medium term, was dead in the water (see Valdmanis 2017).

Without the CPP, the US wind market has to rely on the RPS (Renewable Portfolio Standard) on the State level for support. In addition, tax reforms and border taxes being proposed by the Trump administration could curb the availability of tax equity, increase costs and eventually hurt project economics.

Taking a longer view, Trump's apparent hostility to action against climate change and championing of fossil fuels is unlikely to make a major dent in the global progress being made by renewables. Major countries including China and even fossil-fuel producing countries like Saudi Arabia were quick to point to the strong commitment of the international community to combatting climate change at COP22 in Morocco in the wake of Trump's election, while analysts have predicted that Trump's measures to cut US funding for renewables-related R&D will simply lead to China and other countries such as Germany winning a bigger share of the global market for clean technology goods and services.

Trump's plans to revive the US's coal-generation sector, meanwhile are unlikely to prosper due to the shifting economics of power generation. Coal is declining due to strong competition from cheap natural gas and renewables and given the direction of travel for power costs – analysts expect wind and solar to be significantly cheaper than coal in the 2020s – it is unlikely that utilities and banks will be lining up to build new coal projects. Added to this of course is uncertainty as to how long Trump and his cohorts are likely to remain in power.

The wind industry is far more able to stand on its own two feet than in previous eras, and in confident mood with the PTC due to expire in 2019 in any case. There is strong support for renewables at the state level with almost 30 states having Renewable Portfolio Standards (RPS) and these will continue to incentivise renewable energy production even with the removal of federal support.

Broad public awareness and support, coupled with an increasingly competitive consumer offering thanks to a falling LCOE across multiple renewables categories, have made the industry incrementally less reliant on government support of any kind.

(Goffin, 2016)

Conclusion

Who will reap the wind?

As we have seen, the story of the wind-power sector is one of remarkable growth on a global scale, punctuated by slowdowns and politically led stop–go cycles at the national level.

When I wrote the first edition of this book, the wind industry was suffering a collective crisis of confidence in the aftermath of the failed Copenhagen climate talks and leading companies in the sector seemed on the edge of going out of business.

The world seemed firmly on the road to an unsustainable increase in global temperatures and politicians seemed incapable of creating stable international frameworks that would force a big enough change in investment patterns to reverse the trend. Incumbent interests were looking difficult.

Fast forward a few years and we are in a different world.

In April 2017, the UK saw its first coal-free day since the 1880s and everywhere around the globe, coal retirements sped up. Carbon emissions showed a more positive than expected trend.

Politically, the consensus around the need for concerted international action against climate change was stronger in the aftermath of the historic Paris agreement in late 2015, and this consensus was proving highly resilient despite the decision by the Trump Administration to withdraw from the Agreement in June 2017.

Significantly, leadership on climate and renewables is coming largely from a former climate 'villain' China, which is now arguably leading in every area of clean technology. Other countries that were seen as having the potential to punch a hole through emissions targets because of rapidly increasing coal-power generation, such as India are stepping up their renewables deployment radically and slowing down planned coal deployment. Even the oil exporting countries of the Middle East are moving fast to deploy wind and solar power.

As we saw in a number of places in this book, what is underpinning the steady expansion of wind power is no longer politics, but economics, as the long sought for 'grid parity' is becoming a reality in growing areas of the world. In the words of Mainstream Renewable Power CEO Eddie O'Connor, wind power has reached its 'iPhone moment' (O'Connor 2017).

Public support for wind power is stronger than ever – especially among young people, despite moves by some politicians to further special interests by attacking the technology with spurious arguments.

And in contrast to a few years ago, the use of wind-power technology is embedded in the corporate culture of many of the biggest and most respected brands in the consumer market place. As many commentators have pointed out, Google and Apple have more influence over hearts and minds than governments; and their message is that wind and solar are the future.

There are still big political challenges to be surmounted. European markets still need reforming and redesigning to speed up the replacement of existing fossil-fuel capacity with renewables in a way that makes business sense. Short-sighted politicians can still cause significant disruption to market development, whether it is a small but vociferous group of UK Conservative MPs who dislike wind turbines, or Donald Trump and a coterie of advisors linked to fossil-fuel interests who cannot be underestimated. But in general, the direction of travel is clear.

As Eddie O'Connor said in a major article in the *Financial Times*: 'Fossil fuels have lost. The rest of the world just doesn't know it yet.' (Clark, 2017b).

With power comes responsibility. As wind power and other renewables account for a greater and greater proportion of countries' power mixes, companies are realising that they need to provide credible answers around their impact on the overall system – a so-called 'whole system' approach, and come up with the right policy initiatives to manage transition.

For both Mainstream CEO and Founder Eddie O'Connor and wind-turbine technology pioneer Henrik Stiesdal, the single biggest challenge facing the wind-turbine sector is the variability or intermittency of wind-energy production, which as Stiesdal points out 'is more and more used as an argument to constraint further development in countries and regions with high penetration'.

Moving beyond the constraints will require a broader grid, widespread demand-side management and the development of competitively priced storage. 'Using these three strategies, wind and solar can provide 100% of electricity demand worldwide with zero pollution', O'Connor predicts.

Of course, variability is not the only issue. As we have seen, Solar PV is emerging as a formidable competitor in many markets and wind-energy companies will have to both cooperate and compete with solar in order to continue to thrive.

The NIMBY (Not in My Back Yard) phenomenon and wider concerns around the visual and environmental impact of wind farms will continue to be an important issue for the wind industry – although this is more a market-shaping factor rather than a fundamental challenge.

Meanwhile, wind-turbine companies cannot afford to stand still. As we predicted in the first edition of this book, how the industry evolved was going to depend on the speed of market growth. An uneven, stuttering level of growth was likely to lead to the survival of niche players concentrated in local or regional markets. Slower growth would also mean a continuing level of caution in making investments, and hence a relatively slower growth in productivity.

On the other hand a renewed spurt of sustained growth was likely to spur a long-awaited wave of merger and acquisitions activity and consolidation, and this is exactly what has happened. Today's market is characterised by higher volumes, but lower prices and margins, and this, along with further erosion of national barriers and the globalisation of the supply chain is creating more not less competition in the market.

The wave of consolidation has created a group of front-runners consisting of Gamesa-Siemens, Vestas, and GE who are able to turn their global industrial footprints and strengths in knowhow and R&D into a price advantage. There are also two Chinese players, Goldwind and Envision, who have increased their market shares, own their own intellectual property and are equipped to expand internationally. Others, such Nordex-Acciona or Enercon have their own unique strengths, but in general the wind market will not be a comfortable environment for the smaller players.

The question of which of the countries and economies will eventually benefit the most from the creation of a mature, truly global industry is still an open one.

China is making the big investments in large-scale equipment manufacturing and enabling technologies like grid, combined with a stable policy framework and long-term planning that will prepare it for leadership for decades to come. The US still has formidable strengths in innovation as can be seen in the coming together of West Coast technology disruption and manufacturing in the digital-isation of the wind industry. It is to be hoped that political disruption to the US's wind industry caused by the short-term thinking of the Trump administration will prove to be a bump in the road rather than a major setback – as the US economy has much to lose from abdicating leadership in such a strategic sector. Meanwhile, Europe has created the most powerful and dynamic companies in the wind sector, and these form a key part in the continent's bid to continue to be a relevant industrial power. The likes of Vestas and Siemens-Gamesa continue to play a historic role in spreading the wind industry and the knowhow and industrial culture that goes with it to new countries across the world.

The degree to which markets and policy frameworks facilitate or hinder the development of the industry going forward depends on our ability to convince policy makers to grasp the opportunity that wind power represents. I hope this new edition of *Wind Power* can make a small contribution to this aim.

Meanwhile, the companies that are successful in this new period will be those that are able to think beyond the old model of simply supplying equipment or building wind farms. As FTI Consulting's Aris Karcanias says, 'Those companies with a holistic view betting on information-rich models with a systems approach, that integrate the established technologies such as wind and solar with emerging enablers such as storage and digital solutions will be the best placed to reap the rewards in the energy world of tomorrow.'

The new world of information has yet to play out, and the outlines of the new business models are only just starting to emerge. What is most important is a willingness to continue to innovate and experiment, just as Henrik Stiesdal and his father did in their Danish back garden forty years ago.

References

Backwell, B. (2009a) Sector is hopeful, but warns of 'unravelling' if talks fail, *Recharge*, 19 November, 2009. Available online at: www.rechargenews.com/magazine/article 1284642.ece.

Backwell, B. (2009b) UTC steps in to save Clipper from meltdown, *Recharge*, 17 December 2009. Available online at: www.rechargenews.com/magazine/article 1283348.ece.

Backwell, B. (2010a) Iberdrola chairman flies in as companies seek agreement, *Recharge*, 9 September 2010. Available online at: www.rechargenews.com/magazine/article 1288037.ece.

Backwell, B. (2010b) Tanti outlines strategy to become world number one, *Recharge*, 8 October 2010. Available online at: www.rechargenews.com/news/policy_market/ article1288358.ece.

Backwell, B. (2010c) Is Siemens line the future of turbine production?, *Recharge*, 23 July 2010. Available online at: www.rechargenews.com/wind/article1286518.ece.

Backwell, B. (2011a) Indian court deals Enercon a blow over turbine patents, *Recharge*, 4 February 2011. Available online at: www.rechargenews.com/news/policy_market/ article1289702.ece.

Backwell, B. (2011b) Gamesa seals $2bn turbine coup with India's Caparo, *Recharge*, 17 May 2011. Available online at: www.rechargenews.com/wind/ article1291277.ece.

Backwell, B. (2011c) Indian wind market on course to break 3GW barrier this year, *Recharge*, 25 November 2011. Available online at: www.rechargenews.com/wind/ article1294887.ece.

Backwell, B. (2012a) Blood on the carpet as Vestas restructures, *Recharge*, 19 January 2012. Available online at: www.rechargenews.com/wind/article1295345. ece.

Backwell, B. (2012b) Vestas in disarray: deputy CEO resigns ahead of results, *Recharge*, 7 February 2012. Available online at: www.rechargenews.com/wind/article1299018.ece.

Backwell, B. (2012c) Vestas says former CFO did India deals without board approval, *Recharge*, 2 October 2012. Available online at: www.rechargenews.com/wind/ article1298501.ece.

Backwell, B. (2013a) Interview: UB Reddy, India's Wippa, *Recharge*, 5 July 2013. Available online at: www.rechargenews.com/wind/asia_australia/article1330668. ece.

Backwell, B. (2013b) Brazil: The soft superpower, *Recharge*, 2 September 2013. Available online at: www.rechargenews.com/ThoughtLeaders/article1335038.ece.

Backwell, B. (2013c) Alstom: Brazil turbine costs will fall, *Recharge*, 12 September 2013. Available online at: www.rechargenews.com/wind/americas/article1337072.ece.

Backwell, B. (2013d) Gamesa expects Brazil shake out, *Recharge*, 11 September 2013. Available online at: www.rechargenews.com/wind/americas/article1337040.ece.

Backwell, B. (2013e) Vestas-Mitsubishi in pole position, *Recharge*, 1 November 2013. Available online at: www.rechargenews.com/wind/offshore/article1341789.ece.

Backwell, B. (2013f) GE prefers growth to acquisition, *Recharge*, 5 February 2013. Available online at: www.rechargenews.com/wind/europe_africa/article1316145.ece.

Backwell, B. (2013g) Vestas fraud probe centres around 'ghost' Chinese turbine purchase, *Recharge*, 28 May 2013. Available online at: www.rechargenews.com/wind/article1327963.ece.

Backwell, B. (2013h) Vestas momentum grows – analysts, *Recharge*, 3 July 2013. Available online at: www.rechargenews.com/wind/europe_africa/article1331380.ece.

Backwell, B. (2013i) MAKE: Orders bode well for 2014, *Recharge*, 26 September 2013. Available online at: www.rechargenews.com/wind/article1338759.ece.

Backwell, B. (2013j) Suzlon loses India wind top spot, *Recharge*, 10 May 2013. Available online at: www.rechargenews.com/wind/asia_australia/article1326241.ece.

Backwell, B. and Publicover, B. (2013) The future for foreign firms in China, *Recharge*, 4 October 2013. Available online at: www.rechargenews.com/magazine/article1338688.ece.

Bautzer, T. (2017) Brazil's Light authorizes Renova to enter exclusive talks with Brookfield, *Reuters*, 20 July 2017. Available online at: www.reuters.com/article/us-renova-energia-m-a-brookfield-asset-idUSKBN1A42R2

Burkhardt, P. and Cohen, M. (2016) 'Rogue' power firm threatens fastest renewable expansion, *Bloomberg*, 6 December 2016. Available online at: www.bloomberg.com/news/articles/2016-12-05/-rogue-power-firm-threatens-world-s-fastest-renewable-expansion

Clark, P. (2016) China's Goldwind becomes world's largest wind turbine maker, *Financial Times*, 23 February 2016. Available online at: www.ft.com/content/123f1af0-d97e-11e5-a72f-1e7744c66818

Clark, P. (2017a) Dong Energy breaks subsidy link with new offshore wind farms, *Financial Times*, 14 April 2017. Available online at: www.ft.com/content/f5b164a6-20f8-11e7-b7d3-163f5a7f229c

Clark, P. (2017b) The Big Green Bang: How renewable energy became unstoppable, *Financial Times*, 19 May 2017. Available online at: www.ft.com/content/44ed7e90-3960-11e7-ac89-b01cc67cfeec

Clover, P., Zhao, F. and Backwell, B. (2015) Nordex: The acquisition of AWP shows compelling strategic logic at a good price, *FTI Intelligence Spark Note*, 9 October 2015). Available online at: www.fti-intelligencestore.com/index.php?route=spark/main

Costa da Paula, L. (2015). Bioenergy pede desistência de 547 MW em 19 parques eólicos, *Exame*, 11 June 2015.

CSIR (2017) Formal comments on the Integrated Resource Plan (IRP) Update Assumptions, Base Case and Observations 2016 [20170331-CSIR-EC-ESPO-REP-DOE-1.1A Rev 1.1], 4 April 2017. Available online at: www.csir.co.za/sites/default/files/Documents/IRP_Update_Assumptions_1904.pdf).

Davidson, R. (2016) How Vestas won the Mid West, *Windpower Monthly*, 29 July 2016. Available online at: www.windpowermonthly.com/article/1403499/vestas-won-mid west

Davidson, R. (2017) Goldwind project financing 'a significant milestone', *Windpower Monthly*, 9 May 2017. Available online at: www.windpowermonthly.com/article/1432921/goldwind-project-financing-a-significant-milestone

De Clercq, G. (2013) European utilities CEOs urge end to renewables subsidies, *Reuters*, 11 October 2013. Available online at: http://uk.reuters.com/ article/2013/ 10/11/utilities-renewables-ceos-idUKL6N0I106Y20131011.

Dehlsen, J. (2003) Interview US Department of Energy. Available online at: www.windpoweringamerica.gov/filter_detail.asp?itemid=683.

Dezem, V. (2016). Brazil wind at risk as weak economy crimps demand, GE says, *BloombergQuint*, 3 July 2016. Available online at: www.bloombergquint.com/business/ 2016/06/30/brazil-wind-threatened-as-weak-economy-crimps-demand-ge-says?utm_ source=bloomberg-menu

EWEA (2010) EWEA 2010 statistics: Offshore and eastern Europe the new growth drivers for wind power in Europe, 31 January 2011. Available online at: www.ewea.org/ news/detail/2011/01/31/ewea-2010-statistics-offshore-and-eastern-europe-new-growth-drivers-for-wind-power-in-europe/.

EWEA (2013b) *EU wind power grows in 2012, but industry challenged in 2013*. Available online at: www.ewea.org/news/detail/2013/02/08/eu-wind-power- grows-in-2012-but-industry-challenged-in-2013/.

EWEA Offshore Conference (2013) *Insight at Thought Leaders VIP brunch*. Available online at: www.ewea.org/offshore2013/wp-content/uploads/ EWEA-Offshore-2013-day-three.pdf.

Financial Times (2016) Siemens and Gamesa to merge wind businesses, June 18, 2016. Available online at: www.ft.com/content/01551aec-34ae-11e6-bda0-04585c31b153

Foster, M. (2016). Japan retains wind FIT as solar is cut, *Windpower Monthly*, 25 September 2016. Available online at: www.windpowermonthly.com/article/1385084/ japan-retains-wind-fit-solar-cut.

FTI Intelligence. (2015) Global Wind Supply Chain Update 2015, 12 January 2015. Available online at: www.fticonsulting.com/fti-intelligence/energy/research/clean-energy/global-wind-supply-chain-update-2015

Goffin, R., Mohr, N., Pearson, L. and Clover, R. (2016) Renewables, Trumped: Implications for the U.S. energy sector, *FTI Intelligence Spark Note*, 14 December 2016. Available online at: www.fti-intelligencestore.com/index.php?route=spark/main

González, A., Hübner, A. and De Clercq, G. (2016). Areva rejects GE offer for Adwen JV offshore wind assets – sources, *Reuters*, 31 August 2016. Available online at: www.reuters.com/article/us-gamesa-m-a-siemens-areva-idUSKCN1152EG?type=com panyNews

GWEC (2009) Renewable energy financing in a year of economic turmoil, *Global Wind 2009 Report*. Available online at: http://gwec.net/wp-content/uploads/2012/06/ GWEC_Global_Wind_2009_Report_LOWRES_15th.-Apr..pdf.

GWEC (2011) *Global Wind Report 2011*. Available online at: http://gwec.net/wp-content/uploads/2012/06/Annual_report_2011_lowres.pdf.

GWEC (2012) *Analysis of the regulatory framework for wind power generation in Brazil*. Available online at: http://gwec.net/wp-content/uploads/2012/06/Brazil_report_2011. pdf.

GWEC and IWTMA. (2016) India Wind Market: A Brief Outlook, 2016. [Summary Report]. Available online at: www.gwec.net/wp-content/uploads/vip/GWEC_IWEO_ 2016_LR.pdf

Hoel, A. (2004) A wind Bonus for Siemens, *Power Engineering International*, 1 November 2004. Available online at: www.powerengineeringint.com/ articles/print/volume-12/issue-11/regulars/news-analysis/a-wind-bonus-for- siemens.html.

Hurst, D. (2017). Victory for Japanese nuclear industry as high court quashes injunction,

The Guardian, 28 March 2017. Available online at: www.theguardian.com/environ ment/2017/mar/28/japanese-nuclear-industry-court-injuction-takahama-greenpeace

Karcanias, A., Backwell, B., Zhao, F. and Clover, R. (2016) Siemens' acquisition of Gamesa would create global turbine giant, *FTI Intelligence Spark Note*, 29 January 2016. Available online at: www.fti-intelligencestore.com/index.php?route=spark/main

Kessler, R. (2012) New Clipper owner shrinks workforce, looks to O&M, *Recharge*, 21 August 2012. Available online at: www.rechargenews.com/wind/article1298182.ece.

Lazard. (2016) Lazard's levelized cost of energy analysis – v. 10.0, December 2016. Available online at: www.lazard.com/media/438038/levelized-cost-of-energy-v100.pdf

Leal, M. (2013a) Tolmasquim: 'It's wind's moment', *Recharge*, 2 September 2013. Available online at: www.rechargenews.com/wind/americas/article1334827.ece.

Leal, M. (2013b) Renova and the gold mine, *Recharge*, 3 September 2013. Available online at: www.rechargenews.com/wind/americas/article1335050.ece.

Lee, A. (2013a) Euro CEOs blast support for RE, *Recharge*, 11 October 2013. Available online at: www.rechargenews.com/wind/article1340231.ece.

Lee, A . (2013b) Suzlon agrees debt restructuring, *Recharge*, 24 January 2013.Available online at: www.rechargenews.com/wind/asia_australia/article1315033.ece.

Lee, A. (2014a) Siemens plants UK 'game changer', *Recharge*, 25 March 2014. Available online at: www.rechargenews.com/wind/article1356182.ece.

Lee, A. (2014b) Vestas EBIT beats expectations, *Recharge*, 3 February 2014. Available online at: www.rechargenews.com/wind/europe_africa/article1351179.ece.

Lee, A. and Jensen, K. (2012) Vestas seeks investor for up to 20% stake says chairman, *Recharge*, 24 August 2012. Available online at: www.rechargenews.com/wind/article1298204.ece.

Lee, A. and Publicover, B. (2014) Sinovel faces China criminal probe, *Recharge*, 13 January 2014. Available online at: www.rechargenews.com/wind/asia_australia/article1348792.ece.

Li, J. *et al.* (2012) *China Wind Energy Outlook 2012*. Available online at: www.gwec.net/wp-content/uploads/2012/11/China-Outlook-2012-EN.pdf.

Magee, D. (2009) *Jeff Immelt and the New GE Way*, McGraw Hill.

Mexia, A. (2011) *EDP-CTG Strategic Partnership Establishment Friday, 23rd December*. Available online at: www.edp.pt/en/Investidores/EDP%20Ficheiros/Transcript%20EDP-CTG%20Partnership.pdf.

Milne, R. (2012) Vestas investors seek wind of change, *ft.com*, 2 September 2012. Available online at: www.ft.com/cms/s/0/f4622308-f4dc-11e1-b120-00144feabdc0.html#axzz31lzk2Hzg.

Milne, R. (2013) Vestas eyes 'future without surprises', *ft.com*, 21 August 2013. Available online at: www.ft.com/cms/s/0/641a1190-0a78-11e3-9cec-00144feabdc0.html#axzz31lzk2Hzg.

Mumma, C. (2002) GE seeks refund from Enron Wind, *Bloomberg News*, 15 November 2002. Available online at: http://articles.latimes.com/2002/nov/15/business/fi-wind15.

Nicola, S. and Bauerova, L. (2014) Dirtiest coal's rebirth in Europe flattens medieval towns, *Bloomberg*, 6 January 2014. Available online at: www.bloomberg.com/news/2014-01-06/dirtiest-coal-s-rebirth-in-europe-flattens-medieval-towns.html.

Nordex (2012) *2012 Annual Report Nordex SE*. Available online at: www.nordex- online.com/fileadmin/MEDIA/Geschaeftsberichte/Nordex_GB2012_en.pdf.

O'Connor, E. (2016a) Cost of Eskom Folly is R100bn more a year, *Business Day*, 8 December 2016. Available online at: www.businesslive.co.za/bd/opinion/2016-12-08-cost-of-eskom-folly-is-r100bn-more-a-year/

O'Connor, E. (2016b) Unprecedented victory renewables over fossil fuels in Chile, *Huffington Post*, 31 August 2016. Available online at: www.huffingtonpost.co.uk/dr-eddie-oconnor/post_12955_b_11760070.html

O'Connor, E. (2017) Renewable energy's iPhone moment, *Huffington Post*, 17 January 2017. Available online at: www.huffingtonpost.co.uk/dr-eddie-oconnor/renewable-energys-iphone-_b_14218140.html

Parkinson, G. (2016) New low for wind energy costs: Morocco tender averages $US30/MWh, *RenewEconomy*, 17 January, 2016. Available online at: http://reneweconomy.com.au/new-low-for-wind-energy-costs-morocco-tender-averages-us30mwh-81108/

Patton, D. (2013) The rise and stall of Sinovel, *Recharge*, 6 January 2013. Available online at: www.rechargenews.com/magazine/article1302268.ece.

Ramesh, M. (2012) Vestas to scale down India operations, *The Hindu*, 25 September 2012. Available online at: www.thehindubusinessline.com/companies/vestas-to-scale-down-india-operations/article3935285.ece.

Randall, T. (2016). World energy hits a turning point: Solar that's cheaper than wind, *Bloomberg*, 15 December 2016. Available online at: www.bloomberg.com/news/articles/2016-12-15/world-energy-hits-a-turning-point-solar-that-s-cheaper-than-wind

Recharge (2013a) The Rise and Stall of Sinovel, *Recharge*, 6 January 2013. Available online at: www.rechargenews.com/magazine/article1302268.ece.

Recharge (2013b) Suzlon woes 'dragging on Repower', *Recharge*, 14 August 2013. Available online at: www.rechargenews.com/wind/europe_africa/article1334457.ece.

Renewable Energy in Denmark [Vedvarende Energi i Danmark]. *Tvindkraft*, OVE's Forlag. Available online at: www.tvindkraft.dk/

Sampathkumar, M. (2017) Donald Trump says 'something could happen with the Paris Agreement', *The Independent*, 13 July 2017. Available online at: www.independent.co.uk/news/world/americas/us-politics/trump-paris-agreement-macron-france-visit-climate-change-something-could-happen-a7840021.html

Sawyer, S. (2016). China wind power blows past EU – Global wind statistics release, GWEC, 10 February 2016. Available online at: www.gwec.net/china-wind-power-blows-past-eu-global-wind-statistics-release/

Shankleman, J., Parkin, B. and Hirtenstein, A. (2017) Gigantic wind turbines signal era of subsidy-free green power, *Bloomberg*, 21 April 2017. Available online at: www.bloomberg.com/news/articles/2017-04-20/gigantic-wind-turbines-signal-era-of-subsidy-free-green-power

Shepherd, C. and Hornby, L. (2016) Wind turbine maker Goldwind looks beyond Xinjiang, *Financial Times*, 21 June, 2016. Available online at: www.ft.com/content/6b536324-f1ea-11e5-9f20-c3a047354386

Siemens Gamesa. (2017). [Press release]. 9 May 2017. Available online at: www.gamesacorp.com/en/communication/news/siemens-gamesa-names-markus-tacke-as-new-ceo.html?idCategoria=0&fechaDesde=&especifica=0&texto=&idSeccion=0&fechaHasta=

Smalley, J. (2015). Mitsubishi floating 7MW turbine ready for deployment, *Windpower Engineering & Development*, 29 June 2015. Available online at: www.windpowerengineering.com/construction/projects/mitsubishi-floating-7-mw-turbine-ready-for-deployment/

Stromsta, K.E. (2009) Suzlon asks the billion-dollar question, *Recharge*, 17 December 2009. Available online at: www.rechargenews.com/magazine/article1283355.ece.

Stromsta, K.E. (2013) Interview: Ignacio Martín, Gamesa, *Recharge*, 7 June 2013. Available online at: www.rechargenews.com/magazine/article1327930.ece.

Suzuki, T. (2017). Six years after Fukushima, much of Japan has lost faith in nuclear power, *The Conversation*, 9 March 2017. Available online at: http://theconversation.com/six-years-after-fukushima-much-of-japan-has-lost-faith-in-nuclear-power-73042

Tenddulkar, S. (2017) India to auction 4GW by March 2018, *Windpower Monthly*, 8 May 2017. Available online at: www.windpowermonthly.com/article/1432767/india-plans-auction-4gw-march-2018

Thomson, E., Lombrana, L. M., and Dezem, V. (2016) Chile sees surge in wind power, rattles incumbents, *Bloomberg*, 22 August 2016. Available online at: www.bloomberg.com/news/articles/2016-08-17/chile-power-auction-rattles-market-for-incumbent-electric-firms

Toplensky, R. (2017) GE gets EU greenlight for LM Wind Power deal, *Financial Times*, 20 March, 2017. Available online at: www.ft.com/content/79795d99-f18b-397a-8aed-5f63bcf98630

Valdmanis, R. (2017) States challenge Trump over Clean Power Plan, *Reuters*, 6 April 2017. Available online at: www.scientificamerican.com/article/states-challenge-trump-over-clean-power-plan/

Vecchiatto, P. (2017). South Africa nuclear plans stalled as court rules process unlawful, *Bloomberg*, 26 April 2017.

Vestas (2004) Press release 8 July 2004. Available online at: www.vestas.com/files/Filer/EN/Press_releases/VWS/2004/080704-UK.pdf.

Vidal, C. (2017) Recuperar la credibilidad perdida: la misión de José Luis Blanco como nuevo 'capo' de Nordex, *Bolsamania*, 29 March, 2017. Available online at: www.bolsamania.com/noticias/empresas/acciona-jose-luis-blanco-nordex—2593247.html

Weston, D. (2015) Envision acquires 600MW in Mexico, *Windpower Monthly*, 14 October 2015. Available online at: www.windpowermonthly.com/article/1368413/envision-acquires-600mw-mexico

Weston, D. (2016a) Brazil cancels reserve power auction, *Windpower Monthly*, 15 December 2016. Available online at: www.windpowermonthly.com/article/1418989/brazil-cancels-reserve-power-auction

Weston, D. (2016b) Envision acquires French portfolio, *Windpower Monthly*, 6 December 2016. Available online at: www.windpowermonthly.com/article/1417846/envision-acquires-french-portfolio

Weston, D. (2016c) Nordex and Acciona complete merger, *Windpower Monthly*, 4 April 2016. Available online at: www.windpowermonthly.com/article/1389823/nordex-acciona-complete-merger

Weston, D. (2016d) Siemens and Duke form servicing co-op, *Windpower Monthly*, 3 November 2016. Available online at: www.windpowermonthly.com/article/1414434/siemens-duke-form-servicing-co-op

Weston, D. (2016e). Analysis: Siemens seeks strength in Gamesa's emerging markets, *Windpower Monthly*, 29 February 2016. Available online at: www.windpowermonthly.com/article/1384665/analysis-siemens-seeks-strength-gamesas-emerging-markets

Weston, D. (2017a) Goldwind acquires 530MW Australian site, *Windpower Monthly*, 8 May 2017. Available online at: www.windpowermonthly.com/article/1432734/goldwind-acquires-530mw-australian-site

Weston, D. (2017b) Nordex CEO quits after poor forecast, *Windpower Monthly*, 20 March 2017. Available online at: www.windpowermonthly.com/article/1427875/nordex-ceo-quits-poor-forecast

Williams, M. and Menon, M. (2013) Bewildering Indian policies fuel needless coal imports, *Reuters*, 13 October 2013. Available online at: www.reuters.com/article/2013/10/13/us-india-coal-idUSBRE99C0CB20131013.

Windpower Monthly (2001) Two wind giants go head to head – Vestas and Gamesa split, 1 January 2002. Available online at: www.windpowermonthly.com/article/950913/two-wind-giants-go-head-head—-vestas-gamesa-split.

Windpower Monthly (2004) Danish wind giants head for merger – Vestas and NEG Micon going for global supremacy. 1 January 2004. Available online at: www.windpower-monthly.com/article/956160/danish-wind-giants-head-merger—-vestas-neg-micon-going-global-supremacy.

Yergin, D. (2011) *The Quest: Energy, Security and the Remaking of the Modern World*, Allen Lane.

Index